SEAWATER DESALINATION
Impacts of Brine and Chemical Discharge on the Marine Environment

Sabine Lattemann

Thomas Höpner

Institute for Chemistry and Biology
of the Marine Environment (ICBM)
University of Oldenburg, Germany

Balaban Desalination Publications

SEAWATER DESALINATION
Impacts of Brine and Chemical Discharge on the Marine Environment
by Sabine Lattemann and Thomas Höpner

Published by
Desalination Publications
Via Antica Arischia, 1
L'Aquila 67100
Italy
Tel. +39 0862 3475308
Fax: +39 0862 3475213
E-mail: balaban@desline.com
Web site: http://www.desline.com

Distributed by
DEStech Publications
561 Springton Way
Lancaster, PA 17601-1, USA
Tel. +1 717 290 1660
Fax. +1 717 509 6100
E-mail: aderaco@destechpub.com
Web site: http://www.destechpub.com

ISBN 0-86689-062-9

Contents

1

Introduction

It is evident that future population growth will increase the demand for drinking water. In addition, ongoing industrialization, land irrigation and higher living standards cause a rise in per capita consumption of fresh water. This development is concentrated in a narrow coastal zone, where by today about half the world population lives, which will probably rise to three quarters by 2020 (Agenda 21, 1992). However, ground waters are already excessively exploited and at risk of saltwater intrusion in coastal areas, while surface waters are either scarce or prone to pollution. As good quality raw water becomes less available, pretreatment methods must cope with this problem by purifying water for domestic and industrial purposes. The same technology enables us to use seawater as a source of drinking water, thereby extending our use of marine resources for human benefit. Desalination is already tremendously important in many arid countries, and it has to be expected that it will consolidate its position to satisfy the world's demand for water in the future.

Although desalination removes salts and other contaminants, good raw water quality is still prerequisite for process control. On the other hand, coastal development involves land-based pollution of the sea, among which desalination effluents are causing concern in areas of large installed capacities like the Arabian Gulf (UNEP, 1999; Al-Awadhi, 1999; Subba Rao and Al-Yamani, 2000). In this semi-enclosed sea, about 50 % of the world's seawater plants are located (Fig. A.2 & Fig. A.3 on page 112). Although overall capacities are lower, emphasis should also be given to the Red Sea and Gulf of Aden (UNEP 1997, Fig. A.8 & A.9), the Caribbean or the Mediterranean with regard to future growth of this industry. Irrespective of large-scale developments, impacts from desalination discharge can often be observed on a local level, but depend on the sensitivity of the surrounding ecosystem (Höpner and Windelberg, 1996). Effects can be put down to a changed seawater quality in the mixing zone of the plume, and the best proof is given by the effort of separating outfall and intake structures of a plant. This can be achieved by jetties (Altayaran and Madany, 1992), so that entrainment of discharge water is strictly avoided.

Earlier questionnaires among experts indicate that brine disposal and environmental impact assessment of desalination plants have a high priority for future activities (Herbranson and Suemoto, 1993; Ribiero and Spiewak, 1998). However, only few publications discussed this topic so far, and it seems comprehensible to conclude that human benefit outweighs minor impacts on coastal habitats. For the immensely important product, environmental effects may undoubtedly be accepted. Access to a basic water requirement is a fundamental human right implicitly supported by the Right to Food, Health, Human Well-Being, and Life (Gleick, 1999). However, desalination plants are predominantly built in areas that can afford this costly way of drinking water production. It must therefore be questioned if adverse effects are inevitable from the start, or if they can be mitigated technically by more expensive procedures. Efforts are made to optimize process control, which also restricts degradation of the marine environment. In this context, this work intends to review and summarize existing and available information concerning desalination plant effluents. The focus is on possible marine pollutants in the discharge, which might interfere with the coastal water body including water column, sediments and the inhabitants of this realm.

Structure

A short introduction to the two main desalting processes is considered helpful and emphasis is given to vulnerable points in process design and choice of materials in *chapter 2*. To protect reverse osmosis (RO) and multi-stage flash (MSF) plants from certain types of seawater constituents, chemical pretreatment is required, while cleaning chemicals are used to restore the plant's efficiency periodically. Both procedures are shortly described with regard to the type of foulant.

As similar problems arise in both processes due to the nature of seawater, treatment chemicals are partly alike. However, they are not always identical and basic differences can be attributed to process design. To summarize information, classification into different groups of chemicals was chosen in *chapter 3*, while process-related aspects were subordinated. This chapter aims to describe single chemicals as well as the underlying mechanisms that make them effective additives. In addition, dosage and discharge concentrations of substances will be identified preparatory to chapter 4.

After having discussed why and how chemicals are implemented in desalination plants, their effect on the marine environment will be assessed in *chapter 4* by considering water quality criteria, toxicological information and environmental fate. Properties of seawater like salinity and temperature are also influenced by the process and determine the plume's density, which is in turn important for mixing processes and dilution of residual chemicals. Simplified calculations may allow for an estimate of environmental concentrations without exact knowledge of the specific discharge situation, which can then be compared to toxicological information for a preliminary impact assessment.

Attention is given to the Arabian Gulf in *chapter 5*, which discusses major pollutants of desalination effluents in terms of total loads. This approach takes the Gulf's high installed capacity into account and relates it to the surrounding Gulf ecosystem.

Maps of the region and corresponding figures are attached in the *appendix*, following a summary and outlook in *chapter 6*.

2

Desalination Technology

Extracting pure water from saline water is an energy driven process and relies either on a phase change of the solvent or the forced movement of water molecules through a selective membrane. As fresh water is produced the waste stream is increased in salt concentration and discharged back to the sea, enriched with chemicals added to control the process.

The early desalination industry was based on thermal evaporation in the beginning of the last century and distillation is still the predominant process worldwide today. However, operation was fundamentally modified in the 1950s. Since then, several vessels are connected in series with gradually changing temperature and pressure values, which induce boiling without adding extra heat to that stage of the plant. The processes are therefore termed multi-stage flash (MSF) and multi-effect distillation (MED), with MSF being more common today. In the early 1960s the first membranes were implemented in electrodialysis (ED) plants but reverse osmosis (RO) technology soon overtook ED in terms of market share. As membrane separation experienced major drawbacks compared to the more robust distillation process, improved water pretreatment methods and membrane materials were developed in the past. This trend is going to continue and research currently focuses on new techniques like ultra- and nanofiltration to cope with insufficient feedwater quality. Similar yet earlier experience in distillation improved the overall performance through larger, more efficient plants, more resistant construction materials and specially designed water additives for scale, fouling and corrosion control.

The lead of MSF over RO is based on its earlier application and higher experience in process control, but market shares are almost equal today if all feedwater sources are considered: both provide for slightly more than 40 % of total capacities, which is more than 10 million m^3 per day each. However, these figures do not account for underlying trends with regard to different sources of feedwater. MSF is the dominant process over RO for seawater applications worldwide with 70 % and 18 % respectively[a]. This has to be seen in the context of seawater being the most common raw water source of all: about 15 million m^3 potable water are extracted from seawater each day compared to 6.5 million m^3 per day from brackish water sources and a worldwide total capacity of about 25 million m^3 per day[b]. From these figures, it can be concluded that the MSF process is almost entirely used for seawater desalination while RO plants predominately operate on river or brackish water sources.

The uneven market shares of MSF and RO in seawater applications reflect the situation prevailing in the countries of the Middle East. Saudi Arabia has the world's largest seawater desalting capacity with about 4 million m^3/day and most plants are based on MSF technology (Fig. A.2 on page 111). MSF is even more predominant in the UAE and Kuwait which have second and third largest capacities of seawater desalination. Water production is usually coupled to power generation plants in these countries, thereby making use of the waste heat and explaining the lead of MSF technology. As another consequence, most seawater desalination is achieved by large plants that are capable of producing more than 5000 m^3 per day and unit. The year 2000, which was even better than the peak year 1993 in terms of contracted new capacities, will further imbalance market shares. This will again be due to large thermal plants in the Middle East, as MSF production costs decreased further and fossil fuels are still cheap (pers. communication Wangnick, 2001).

In contrast, the RO process is less energy intensive and consequently a favored option in countries which do not have large oil and gas resources at their disposal. However, these capacities are too small to have a marked influence on the overall share. For example, Spain follows Kuwait on 4th position with about 0.8 million m^3/day of drinking water extracted from

[a] Total capacities are about 10 and 2.7 million m^3/day each.

[b] Statistics refer to the end of 1999, based on El-Dessouky (2000) and Wangnick (1999).

seawater. This is about two thirds of Spain's total capacity, which was doubled during the last 5 years and is based to 84 % on RO while MSF contributes only 5 % (El-Dessouky, 2000; Wangnick, 1999). The increasing demand in Mediterranean countries of the EU will allow seawater desalination and probably RO to be applied on a larger scale, especially where tourist areas and industrial zones expand locally (Ribiero and Spiewak, 1998). However, RO will remain the predominant process for small, single purpose plants (El-Dessouky, 2000) while MSF is chosen for large dual purpose plants.

As multi-stage flash and reverse osmosis are the prevailing technologies in the field of thermal and membrane desalination, both processes will be shortly introduced in the following sections (2.1 and 2.2 on page 16). A detailed discussion of technology, however, cannot be the purpose of this work and process design of MSF and RO will be considered only so far as necessary in the context of why and how seawater is treated and what the properties of the waste stream are.

2.1 Reverse osmosis

An existing concentration gradient causes a net flow of solvent and dissolved material which finally evens out the disparity. If two compartments are separated by a membrane that is not permeable for dissolved constituents, the solvent will pass the barrier in an osmotic flow. To reverse this osmosis, external pressure can be applied to a concentrated solution like seawater to force water molecules through a membrane. By extracting pure water, the concentration in the rejected brine increases. As a gradient builds up, water diffuses back to dilute the brine so that the external pressure has to be considerably higher than the osmotic pressure to achieve a net flow of product water. The rejected ions and molecules, however, are not dispersed uniformly in the waste stream but accumulate close to the membrane surface, which is termed concentration polarization. This boundary layer even rises energy consumption but more important in the context of this work, it may concentrate salts above their solubility product (Glucina and Laîné, 2000). Precipitation can seriously impair the membrane's performance and furthermore, suspended solids, organic molecules and bacteria can cause irreversible damage if they get caught on the membrane. Therefore, the performance of an RO system depends on pretreatment of the feed water and currently available RO membranes are generally not robust enough to operate directly on seawater (Bou-Hamad et al., 1997).

As preventive methods must suit the type of membrane, basic materials and configurations will be described below (2.1.1) before common types of fouling in seawater reverse osmosis (SWRO) and their corresponding pretreatment and cleaning methods will be considered in sections 2.1.2-2.1.4. This chapter will focus on more general aspects of RO while single additives and their impact on the environment are dealt with in chapters 3 and 4.

2.1.1 Membrane materials and configuration

Materials

The first commercially produced membranes were composed of cellulose acetate (CA) but current CA-membranes are usually blends of cellulose di- and triacetates. Supplementary, polyamide membranes and especially aromatic polyamides were developed and are increasingly used today. This may be put down to their better rejection of dissolved solids, so that drinking water could be produced at lower operating costs (Redondo and Lomax, 1997).

Another reason for this shift is the lower durability of cellulose acetate in aquatic environments and membranes have to be replaced more frequently due to hydrolysis. To reduce decomposition in seawater, plant operation is restricted to pH 4-6.5 whereas polyamide may be exposed to pH 4-11 during operation and an extended pH-range of 2.3-11.9 for short cleaning periods (DuPont, 1994). Furthermore, cellulose acetate may be degraded by enzymatic biological attack but aromatic polyamides are not, although they can be impaired by biological film formation, too. To protect both types of membrane material, a biocide is usually added to the feedwater. A major advantage of CA is that it may be exposed to residual levels of 1 ppm chlorine before oxidation damages the membrane, therefore it can be easily disinfected in surface water with high biological fouling potential (Hydranautics, 2000). Dechlorination is not required for CA but is necessary for polyamide material which is more easily oxidized by residual chlorine in the membrane feedwater.

FIGURE 2.1: Spiral-wound configuration (left) and hollow fiber permeator (far left).

Configuration

The two basic membrane configurations are hollow fiber and spiral-wound (Fig. 2.1). The hollow fiber membrane consist of very small-diameter capillaries (85 μm) which are densely arranged in bundles. In this way, the membrane surface area per element is enlarged, but due to the small spacing among hollow fibers they are very sensitive to the raw water quality in terms of particle trapping (DuPont, 1994; Glucina and Laîné, 2000). Hollow fibers are usually made out of a single polymer material, which can be either cellulose acetate or aromatic polyamide although the latter is more common.

For spiral-wound membranes, organic cellulose acetate may be used too, but synthetic membranes are usually thin film composites (TFC) of different polymers. While polyamide is used for its salt-rejecting property, other materials like polysulfone or polyurea only serve as supporting layers. The membrane is formed into rectangular sheets which are rolled around a collector for product water. This configuration leaves more interstitial space compared to hollow fiber permeators, i.e. several millimeters instead of 25 μm only. While larger spacing makes the permeator more robust against particle trapping, the spiral-wound membranes are thinner than hollow fibers and thereby also more susceptible to damage by particles and erosion. Another major advantage of looser packing are higher flux rates of spiral wound membranes, expressed in liters per day and m^3 element. On the other hand, high flux rates are accompanied by more severe scaling problems due to concentration polarization in the boundary layer (cf. section 2.1 on page 7).

To conclude, both materials and configurations require identical pretreatment although different emphasis is given to levels of residual chlorine, pH range and suspended material. Membrane cleaning is also aimed at removing the same type of foulants but chemicals must again comply with the type of material. The configuration is also important with regard to the cleaning procedure. Spiral-wound membranes cannot be backflushed and flushing can only be done with permeate water to remove deposits (Glucina and Laîné, 2000). Attention must furthermore be given to glues, which can be dissolved by improper chemical treatment. In contrast, hollow fibers require more pretreatment as they are very likely to clog and consequently more difficult to clean (Redondo and Lomax, 1997). Apart from these aspects, the trend is towards synthetic membranes and most membrane suppliers offer either hollow fiber (e.g. Dupont B-10) or spiral-wound configurations (e.g. SWC-types from Hydranautics, Filmtec's FT-30 or Toray's SU-series) for seawater applications.

2.1.2 Conventional seawater pretreatment

As permeators are liable to fouling and added chemicals must suit the type of membrane, most manufacturers offer consultancy for pretreatment and cleaning during membrane warranty periods and provide relevant information in product manuals, e.g. DuPont (1994), Hydranautics (2000) and FilmTec (2000). Among people concerned with desalination it is commonly accepted that pretreatment is necessary for a good plant performance (e.g. Alawadhi, 1997; Al-Shammiri et al., 2000) and chemical consumption contributes up to 25 % to the overall operation cost (Ebrahim et al., 1995). As it is in the interest of the desalting industry to reduce production costs, additives will not be overdosed intentionally. However, the optimum dosage is often difficult to obtain (sometimes by trial and error) and a surplus might ensure safe operation in the case of deteriorating feedwater quality.

The intake seawater can damage the membrane by suspended or dissolved, organic or inorganic matter, so that the term fouling must be used in the broadest possible sense. These parameters are generally more variable in surface water than in water extracted through beachwell intakes, which is naturally filtered and less subject to seasonal changes (cf. section 2.1.3 on page 14). Pretreatment may therefore be extensive or minimal: it might consist of acid addition and cartridge filtration for well water (Redondo and Lomax, 1997), while common steps in surface water purification are scale inhibition, chlorination, in-line coagulation-flocculation, cartridge filtration and dechlorination (Fig. 2.2 on page 17, Ebrahim et al., 1995). This more elaborate 'conventional' procedure is seen in many RO plants of the Middle East today (Winters, 1997) and will be discussed in the following paragraphs with respect to the source of fouling.

Biofouling

Organic matter comprises living or dead particulate material and dissolved molecules. Pretreatment for these components should be considered when total organic carbon (TOC) exceeds 3 mg/l (FilmTec, 2000), which will be the case in most coastal waters. For example, Baig and Kutbi (1998) specified an average of 3.5 ppm TOC in the intake water of the RO plant in Al Jubail, Saudi Arabia. Suspended particles are dealt with in the next paragraph and cartridge filters provide a final barrier for them including most biofouling organisms. However, some bacteria and algae will nevertheless pass this barrier so that the number of living cells has to be reduced prior to filtration. These organisms may otherwise damage membranes by biological attack (CA only, cf. p. 8) or intensive growth (DuPont, 1994). An existing biofilm is rather resistant to chemical cleaning, which stresses the need for adequate pretreatment to avoid film formation from the outset. The organisms surviving pretreatment or cleaning can initiate an extensive regrowth[c] in a nutrient-rich environment. This is provided when dissolved organic matter (DOM) like polysaccharide is rejected by the membrane and accumulates in the boundary layer (concentration polarization). Dissolved organics may not only intensify aftergrowth, but can also settle on the membrane themselves. This type of fouling can be enhanced in the presence of other chemical compounds, for example, humic substances can coagulate with metal hydroxides (Ning, 1999) or pretreatment chemicals (Winters, 1997).

In most cases, a biocide (usually chlorine) is added to the feed water and dosage depends on raw water quality or might even be unnecessary for well water. Chlorine is generally added at the intake so that the plant's piping is protected and to allow for a long enough reaction time. In se-

[c] Also termed aftergrowth as it occurs after chlorination.

vere cases, shock treatments with chlorine or other oxidizing chemicals are carried out. As polyamide membranes are increasingly implemented, most desalination plants have dechlorination facilities installed right after their cartridge filters. Residual chlorine can be depleted by activated carbon filters or sodium bisulfite injection. Chlorine is only effective against living organisms, while organic molecules are either chlorinated or broken down to assimilable nutrients. Concentrations of humic substances, for example, can be reduced by controlled coagulation with hydroxides in the coagulation-flocculation device of the plant. However, chlorination occurs ahead of this section, so that by-products can be formed before the humic material is effectively deposited and removed with other suspended material. Chlorine and other biocides are dealt with in section 3.1 on page 23. Alternatively, ultrafiltration is a non-conventional pretreatment method which will also decrease the concentration of most organic components in the feed stream and is described besides other techniques in section 2.1.3.

Suspended material

Fouling is most likely caused by suspended particles, e.g. clay and silt (per definition $\leq$63 μm), pico- and nanoplankton ($\leq$20 μm), bacteria (1-3 μm) and smallest colloids down to 1 nm particle size (Falbe and Regitz, 1995). Although colloids are by an order of magnitude smaller than most suspended material, they receive special attention as they can coagulate during reverse osmosis and deposit on the membrane surface. Most manufacturers specify an upper Silt Density Index (SDI)[d] of 4-5 to minimize the risk for the permeator (Wilf and Klinko, 1998). Well water is normally below this limit but surface seawater is in the range of 5-40 depending on location (DuPont, 1994).

A 5-10 μm cartridge filter usually precedes the RO membrane as a final protection device for larger particles that might clog the 25 μm spacing of hollow fibers or the larger gaps of spiral-wound membranes. However, most particles are removed ahead of cartridge filters in the same process by which colloids are treated. A conventional practice is by controlled in-line coagulation-flocculation after chlorination has taken place. Coagulants are added to the feedwater and thoroughly mixed so that flocs form which are

[d] A measure for the concentration of suspended material and determined by the rate of plugging of a 0.45 μm filter at 207 kPa.

usually held back by a dual media sand-anthracite filter. Supplementary, pH is controlled by acid addition (Ebrahim et al., 1995) and polyelectrolytes can be added to enhance coagulation-flocculation. The design of surface intake structures, e.g. in deep water and away from fine-grained sediments, can contribute to a reduction of chemical pretreatment. Possible colloid sources are not only clay sediments but also exist inside the plant due to corrosion of tubing material. Although highly resistant materials are used today (stainless steel, i.e. iron with about 20 % chromium, up to 25 % nickel and 6 % molybdenum), small losses due to crevice corrosion may cause membrane fouling (Morton et al., 1996), especially by iron colloids (DuPont, 1994). Coagulating chemicals and the underlying mechanisms of reaction are discussed in detail in section 3.2 on page 27.

Scaling

Membranes are designed to reject dissolved seawater components so that fouling only occurs when dissolved constituents like sparingly soluble salts or metal oxides precipitate and settle on the membranes. Therefore, the composition of total dissolved solids is important. NaCl salt for example is not deposited in desalination plants as the high solubility limit of NaCl is not exceeded in the brine, although sodium and chloride are the major ions in seawater. In contrast, minor ions like calcium, magnesium, strontium and barium may readily form scales if high conversion ratios increase brine concentration[e]. Seawater reverse osmosis (SWRO) is restricted to 20-60 % recovery to limit energy consumption, which is lower than recovery in brackish water reverse osmosis (BWRO) plants. Therefore, solubility limits for sulfate scales ($CaSO_4$, $BaSO_4$ and $SrSO_4$), silicates (SiO_2) and calcium fluoride (CaF_2) are generally not exceeded in seawater brine but these salts may be deposited on brackish water membranes. As calcium carbonate ($CaCO_3$) is very sparingly soluble (Spiegler and El-Sayed, 1994), it is the only salt for which pretreatment is required in SWRO. Precipitation starts at low conversion ratios of only 15 % as calculated by Al-Shammiri and Al-Dawas (1997). Due to relatively low concentrations of Sr^{2+} and Ba^{2+} in seawater, their carbonate scales contribute only insignificant amounts to

[e] A plant which is operated at 50 % conversion produces 1 m^3 of fresh water from 2 m^3 of seawater, so that the brine concentration will be doubled (modern membranes have a nominal salt rejection of 99.7 %, Wilf and Klinko, 1998).

overall deposits (Butt et al., 1997). Calcium carbonate precipitation can be avoided by different additives. Common pretreatment is acid addition, and additionally or alternatively, scaling inhibitors are used. Section 3.3 on page 28 summarizes the different mechanisms and reactions of scale control additives.

Furthermore, iron oxide/hydroxide, aluminum and manganese deposits impair the membrane's performance and can be classified as scales, too. These metals can either be present in seawater or derived from corroding construction material. As iron and manganese are insoluble in their oxidized form (Fe^{3+} and Mn^{4+}), surface seawater will only contain trace amounts of these metals that could be oxidized and deposited during RO. Moreover, Fe^{2+} is usually a limiting nutrient for primary production and likely to be depleted in surface water. In contrast, anoxic well water has a high fouling potential in the presence of oxygen or other oxidants but is not considered a major feedwater source. The feedwater can also be contaminated with corrosion products (Dudley and Darton, 1997), if carbon steel instead of stainless steel and PVC for high and low pressure piping are used as construction materials (DuPont, 1994). Corrosion in seawater can be slowed down if gases (O_2 and CO_2) are forced out of solution and acid treatment is replaced by scaling inhibitors, which will also increase metal solubility. Dissolved metals can also co-precipitate during coagulation-flocculation, but on the other hand, coagulants might be overdosed and this extra material can settle on the membrane, too.

2.1.3 Non-conventional pretreatment

Conventional pretreatment is sometimes regarded as unreliable in terms of constant feedwater quality. In these cases, replacement by beachwells or prefiltration with fine-pored membranes may be possible. However, these methods are not common practice yet and will only briefly be mentioned in this section to point out some recent efforts that could reduce chemical discharge.

Microfilters (MF) remove colloids and suspended matter larger than 0,1 μm while ultrafiltration (UF) even holds back macromolecules down to 0,001 μm (Falbe and Regitz, 1995). More recently, nanofiltration (NF) technology is considered very effective as NF lowers the SDI below 1 similar to MF and UF, but additionally decreases total salinity by up to 50 % in the RO feedwater. As NF membranes are furthermore less permeable for divalent ions, the amount of scale forming ions can be reduced by up to 90 % (Hassan et al., 1998).

It is often claimed that chemical consumption can be reduced considerably if non-conventional methods of pretreatment are used. Additives for coagulation can be omitted (Bou-Hamad et al., 1997) as well as continuous chlorination (Wilf and Klinko, 1998). As NF technology allows for higher recovery rates by lowering RO feed water concentrations, less brine and therefore less chemicals are discharged back to sea per unit freshwater, "making the process more friendly to the marine environment" (Hassan et al., 1998). However, it should be mentioned that the fouling problem is likely to occur on the pretreatment membranes which have to be cleaned periodically. This can be done by backwashing, e.g. every 40 minutes or chemically enhanced cleaning every 6 months (Sheikholeslami, 1999).

2.1.4 Membrane cleaning

Even with the best pretreatment scheme, membranes will eventually foul over a long period of time and chemical cleaning becomes necessary (Redondo and Lomax, 1997). If pretreatment is insufficient, cleaning will not restore the membrane's performance and membranes have to be replaced in the end, which is not acceptable for large plants. In other words: frequency of cleaning directly reflects the performance of the pretreatment system (Chida, 1997) and neither of them can be omitted. However, the most common irreversible damage is due to pretreatment by chlorine and insufficient dechlorination (Dudley and Darton, 1997).

Routine cleaning intervals therefore differ and have to be determined for each plant separately by monitoring salt passage or pressure drop across the membrane, which signal initial fouling. The type of foulant determines the required cleaning solution and manufacturers usually recommend commercial products in addition to chemicals that are effective and compatible with their membranes. Many of these are discussed in section 3.6 on cleaning chemicals, and their possible impact on the marine environment is considered in section 4.8. As intervals, foulants and cleaning agents are all variable, it is impossible to estimate the consumption of certain chemicals. However, the amount of cleaning solution also depends on plant size, independent of its composition and frequency of use. For example,

95 l are necessary per 8 inch permeator bundle (about 20 cm diameter, DuPont, 1994). One of the largest SWRO plants in the world is located in Al Jubail (Saudi Arabia, Arabian Gulf) and consists of 15 units with 205 Dupont B-10 Twin Hollow Fiber permeators each (Baig and Kutbi, 1998). As every twin permeator contains two bundles, a total volume of $2 \cdot 205 \cdot 15 \cdot 95\ l = 584\ m^3$ is at least required for cleaning, which is about 0.6 % of the plant's capacity (91,000 m^3 per day) and 0.3 % of its daily brine discharge[f]. With regard to sparsely available discharge data, the impact assessment of cleaning chemicals in chapter 4 is based on total loads as calculable for the example of Al Jubail. It is not explicitly mentioned that the cleaning solutions are discharged to the sea although it may be assumed and DuPont (1994) advises to meet local discharge standards. Far more frequently than cleaning (e.g. once a day), "backwashing" is necessary to remove deposited material by reversing the flow direction through media filters. Backwashing of dual media filters should be done with sufficiently chlorinated water (FilmTec, 2000) and backwashing water is usually discharged to sea.

2.2 Multi-stage flash

The typical design of an MSF plant consists of a heat rejection section, a heat input section and a heat recovery section in between (Fig. 2.2). The raw seawater is screened and chlorinated before it enters the heat rejection section and the heat recovery section in succeeding stages. The relatively cool seawater is pumped within condenser tubes from one stage to the next, while in turn hot brine streams at the bottom of these stages in the opposite direction. The brine partly flashes into steam upon entering the next stage and condenses on the cool outside of the condenser tubes, thereby heating the incoming seawater within the tubes. As the brine temperature decreases from stage to stage, flashing in the following stage is achieved by superheating the brine at low pressure. In the heat input section of the plant, heating steam is added and condenses on the cool condenser tubes, too. The seawater within the tubes thereby reaches its highest temper-

[f] Baig and Kutbi (1998) specified a seawater flow rate of 12,000 m^3/h for Al Jubail, so that a daily seawater intake of 288,000 m^3 and a brine discharge of 197,000 m^3/day was assumed.

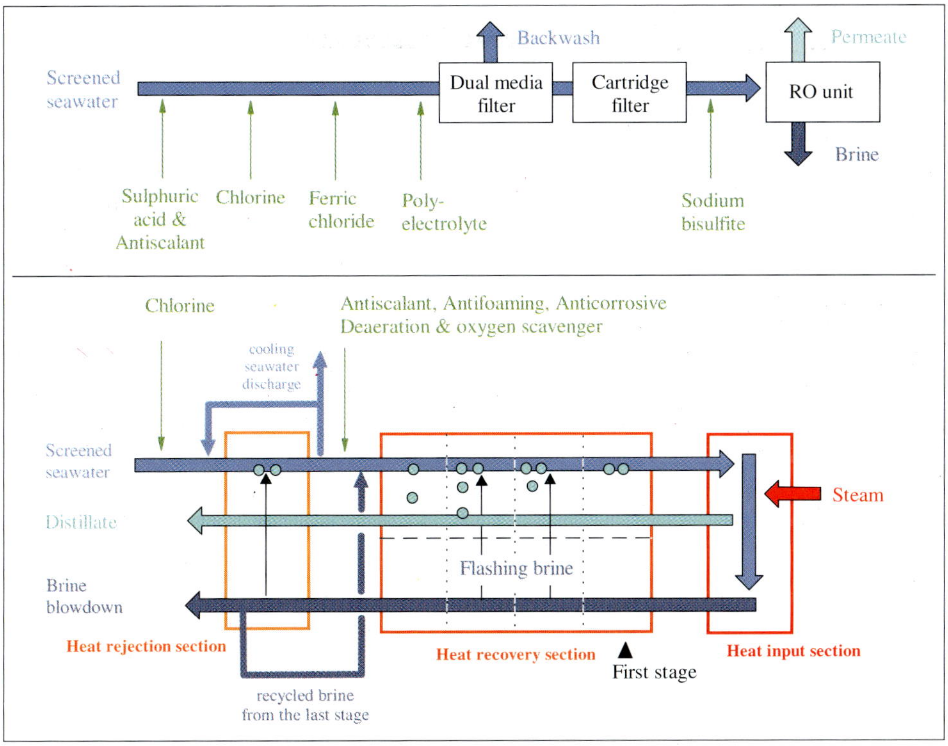

FIGURE 2.2: Reverse osmosis (top) and multi-stage flash (bottom) flow sheets, including the main steps of seawater pretreatment.

ature, the top brine temperature (TBT), and flows now termed as brine back through the succeeding stages. In the last stage, the brine is divided into a recycle stream and a blow-down stream, which is rejected to the sea. The recycle stream is combined with the new feed and reenters the heat recovery section. Partly recycling the brine increases freshwater extraction and simultaneously reduces chemical consumption for pretreatment of the plant (El-Dessouky, 2000). Today, most MSF plants operate with circulating brine but Hamed et al. (1999) state that without recycled brine, less antiscalant is necessary as the brine becomes less concentrated. Between heat rejection and heat recovery section, the incoming water is also divided into a discharge stream and a feed stream. The discharge only serves for cooling and is not chemically treated despite chlorination. Further pretreatment is only done for the feed stream, as described in the following section, and residual chemicals are present in the brine blow-down from the last stage. As the brine is usually blended with twice the amount of reject cooling water (Oldfield and Todd, 1996), salinity, temperature and chemical concentrations will be decreased prior to discharge to the sea.

Due to operation in heated seawater, process design has to take problems into account that arise from this harsh environment and different pretreatment steps are generally necessary to increase the life time of MSF plants (Heitmann, 1990). Corresponding to the part on RO, the following sections are restricted to more general aspects of pretreatment and cleaning in MSF plants. Chemical additives and their impact on the environment are again dealt with in chapters 3 and 4.

2.2.1 Seawater pretreatment

Scaling

Scales are likely to precipitate inside the MSF condenser tubes, thereby lowering heat exchange, promoting corrosion and finally blocking or wearing tubes down. Similarly to RO plants, calcium carbonate may be deposited if high conversion ratios increase the ion concentration in seawater so that solubility products are exceeded. Furthermore, high operating temperatures increase the risk of scaling in MSF plants but the presence of nuclei for crystallization is also critical to initiate scale deposition. These nuclei may be crystals of the same substance or any other solid material found in seawater (Spiegler and El-Sayed, 1994).

Alkaline scales like $CaCO_3$ and $Mg(OH)_2$ become less soluble with increasing temperatures as bicarbonate thermally decomposes and thereby gives way to the scale forming species. Calcium sulfate is also inversely soluble but precipitation starts at higher temperatures (> 115 °C) than observed for alkaline scales and forms very hard scales in MSF plants that are difficult to remove (Buros, 1999). A high TBT is desirable despite scale formation as it allows for more stages to be implemented, thereby increasing the area available for heat transfer and improving the performance ratio (PR)[g] of the plant. In a typical desalination plant, heat transfer areas are

[g] The PR is the ratio of fresh water production to steam input. A typical performance is the production of 8-9 kg distillate from 1 kg steam at a condensation temperature of 73 °C (Heitmann, 1990).

extended by up to 40 % to allow for scaling and biofouling, which counteract and reduce heat exchange across tube walls (Al-Ahmad and Aleem, 1993).

This surplus allowance shows that scaling is of major concern in MSF desalination and needs adequate pretreatment. As a rule, effective high temperatures can only be realized with scale control mechanisms (Gill, 1999), which can be acid addition or antiscalants for alkaline scales. While alkaline scales are relatively easy to control, no commonly accepted treatment exists for sulfate scales. Instead, MSF plants are operated below solubility limits set by temperature and concentration, i.e. below 120 °C and twice the seawater concentration in the brine (Hamed et al., 1999; Patel and Finan, 1999). For example, the last stage before blow-down is prone to scaling although temperatures are relatively low. However, brine salinity is high and should be limited to 70,000 ppm to avoid oversaturation (El-Dessouky, 2000). Operation limits may be exceeded in the future with some inhibitors that are claimed to be effective against sulfates scales (e.g. Gill, 1999; El-Dahan and Hegazy, 2000). A different approach for MSF pretreatment that would allow conversion ratios to be increased is nanofiltration (cf. section 2.1.3). As this process is not very common yet, the emphasis of this work will be on conventional scale control additives, which are discussed in section 3.3 on page 28. In support of antiscalants, on-line ball cleaning systems are highly effective in removing sludge that might form within the tubes. With the implementation of scale control mechanisms, plant shut-down for acid cleaning (p. 21) could be reduced to intervals of once per year (Hamed et al., 1999).

Biofouling

Biofouling in MSF plants starts with the adsorption of dissolved organics to the interior surfaces of heat exchanger tubes and brine channels, where they provide a suitable environment for attachment of film-forming bacteria followed by fungi and protozoa. While this stage of fouling would inevitably result in membrane replacement in RO systems, biofouling inside condenser tubes will predominantly reduce heat exchange and thereby plant efficiency. Plant design compensates for this decrease foresightedly by enlarging surfaces areas. However, lasting biofilms are thought to cause corrosion (Winters, 1995). To control biological fouling, chlorine is added continuously at the intake. Similar to RO plants, chlorine may provide assimilable nutrients by breakdown of larger organics. If these surplus nutrients cause rapid re-growth, shock chlorination may be carried out from time to time. Biocides for MSF and RO plants are discussed in section 3.1 on page 23.

Corrosion

Materials for construction of MSF plants are selected for their ability to withstand the harsh seawater environment, as corrosion is enhanced in high temperature, high salinity water containing residual chlorine and corrosive gases.

Common materials are copper-nickel alloys for their good thermal conductivity as well as their good corrosion resistance, which increases with nickel content. Therefore, Cu-Ni 70-30 alloys are generally used for high temperature stages and the heat rejection section. Although the latter is exposed to temperatures only slightly above ambient, high seawater flow rates transport dissolved oxygen to the material's surface and the high nickel alloy provides adequate protection. In contrast, the cheaper Cu-Ni 90-10 is used in the low temperature stages of the heat recovery section (Oldfield and Todd, 1996; Shams El Din et al., 2000b), where the seawater is deaerated. Trace components further improve longevity, for example manganese and iron can be added at 1-2 % each. However, Cu-Ni-alloys are not totally immune. Observed corrosion rates can remove up to 0.025 mm/a from the surface even if oxygen has been depleted (Todd, 2001), and improper process control or low pH values during acid cleaning can easily increase corrosion.

Other materials are implemented like titanium for heat exchanger tubes or stainless steels for flashing chambers. Both are classified as very resistant and moreover, titanium was found to perform well in acidic environments if corrosion inhibitors are present (Andijani et al., 2000). If stainless steel is used, loss of about 0.11 mm per year must be taken into account although the surface forms a protective layer of Cr_2O_3 (Falbe and Regitz, 1995). As steels for MSF plants commonly contain 17 % chromium, 12 % nickel and 2.5 % molybdenum (Morton et al., 1996), these ions will be released in addition to iron and may be present in the brine. More localized forms of corrosion like pitting and crevice formation are often encountered in desalination plants, too, and increase the average heavy metal content in the brine. Local attack is of concern for stainless steel (Al-Mayouf et al., 1998) as well as for copper-based material and is usually caused by residual oxygen (Oldfield and Todd, 1996) or chlorine levels (Malik et al., 1999).

To conclude, protection of construction materials is achieved by anti-corrosive additives, if an acid wash is conducted, and deaeration of the feed stream to the heat recovery section during normal operation. Depletion of oxygen and carbon dioxide is of great importance, especially as carbon dioxide is increased after acid treatment for scale control. Finally, protective film formation can be enhanced chemically, and all methods will be discussed in section 3.5 on page 38 on corrosion control agents.

Foaming

Seawater contains dissolved organics that accumulate in the surface layer and are responsible for whitecapping under wave action. In desalination plants, a surface film and foaming is not desirable as they increase the risk of salt carry-over and contamination of the distillate. A surface film increases the viscosity at the water-air interface. If flashing is suddenly induced, it is likely that brine droplets burst into the air if the surface tension is too high. In addition, deaeration can be insufficient if gas bubbles are restricted by a strong surface film and cannot escape completely.

To ensure the quality of the distillate and low oxygen levels for corrosion control, antifoaming agents are added to reduce surface tension in the flashing chambers and deaeration section of the plant (cf. section 3.4 on page 37 for antifoaming additives).

2.2.2 Chemical cleaning

Although scale control mechanisms minimize chemical cleaning, a thin film usually builds up inside the condenser tubes and restricts heat transfer. When the performance reaches values below design levels[h], plant shutdown for acid washing becomes necessary which will remove alkaline scales. Sulfate scales must be cleaned off mechanically and deposition should be avoided in the first place. Usually, a pH of 2 is maintained for two hours

[h] For example when the performance ratio falls below 7 but should be around 9 (El-Dessouky, 2000), e.g. once or twice a year (Abdel-Jawad and Al-Tabtabaei, 1999).

in 5000-7000 m^3 warm seawater of 65 °C, and extra acid is added if the pH increases due to reaction with the scale. The cleaning volume depends on distiller size (Shams El Din et al., 1994), but a typical distiller unit produces 30,000 m^3 per day (Morton et al., 1996). An inhibitor is usually added to protect the tubes from corrosion.

MEMBRANE CLEANING SOL'NS - CELLULOSE ACETATE MEMBRANES DO NOT REQUIRE DECHLORINATION B/C THEY CAN HANDLE UP TO 1PPM
POLYAMIDE MEMBRANES DO REQUIRE DECHLOR.

ANTI-SCALING AGENTS

[ ] OF SOURCE-WATER CONSTITUENTS

→ BACKWASH WATER AT LEAST 1x/DAY - WHERE DOES IT GO?

EFFLUENT MODELING WILL GIVE AN IDEA OF CHEMICAL DISPERSAL

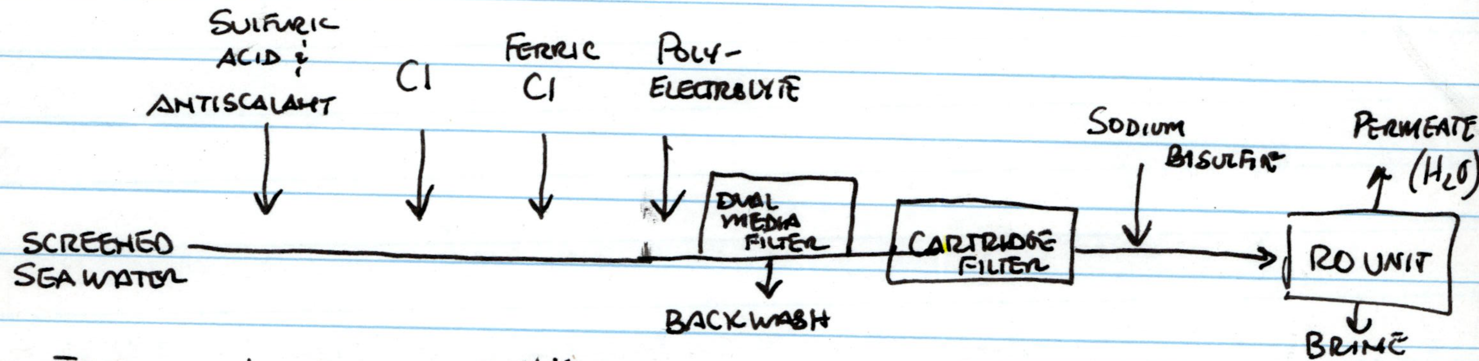

FROM LATTEMANN AND HÖPNER 2003 SEAWATER DESALINATION, IMPACTS OF BRINE AND CHEMICAL DISCHARGE ON THE MARINE ENVIRONMENT

METALS FROM TUBING (Fe, Cr, Ni, Mn = STAINLESS STEEL)

3

Chemical Treatment

3.1 Biocides

Chlorine is commonly used for controlling biological growth in both MSF and RO plants because it inactivates most microorganisms quickly by oxidizing their cellular tissue. It is usually added at the intake as chlorine gas in larger plants, but smaller facilities may alternatively use the hypochlorites of sodium and calcium for easier handling. Irrespective of the initial form, Cl_2, NaOCl and $Ca(OCl)_2$ hydrolyze immediately to hypochlorite (OCl^-) in water:

$$NaOCl + H_2O \rightarrow HOCl + NaOH \quad (3.1)$$
$$Cl_2 + H_2O \rightarrow HOCl + HCl \quad (3.2)$$
$$HOCl \rightleftharpoons H^+ + OCl^- \quad (3.3)$$

However, seawater typically contains 65 ppm bromide, so that hypochlorous acid is displaced by its counterpart, hypobromous acid (HOBr):

$$HOCl + Br^- \rightleftharpoons HOBr + Cl^- \quad (3.4)$$
$$HOBr \rightleftharpoons H^+ + OBr^- \quad (3.5)$$

Free residual chlorine, which is the sum of chlorine gas and all species containing hypochlorite (OCl^-) or hypobromite (OBr^-), is usually between

TABLE 3.1: Biocide concentrations in RO and MSF plants

MSF	Dosing ppm	At outlet ppm	
Chlorine	0.4		Shams El Din and Makkawi (1998)
	1.1	0.25	Shams El Din and Mohammed (1998)
	2	0.2	Burashid (1992)
	2	0.5	Abdel-Jawad and Al-Tabtabaei (1999)
	2 – 4	0.5	Khordagui (1992)
	0.25		Imam et al. (2000)
	0.4 – 0.45		Shams El Din et al. (2000a)
Shock treatment	1		Shams El Din et al. (2000a)
	1		Shams El Din and Makkawi (1998)
	4		Abdel-Jawad and Al-Tabtabaei (1999)
	6		Burashid (1992)
	8		Khordagui (1992)
Hypochlorite	2		Burashid (1992)
RO			
Chlorine	0.5		Abart (1993)
	1.94		Ebrahim et al. (1995)
	2		Morton et al. (1996)
	6		Woodward Clyde Consultants (1991)
	0.5 – 1		Redondo and Lomax (1997)
Sodium bisulfite			
For dechlorination	9		Morton et al. (1996)
	3		Redondo and Lomax (1997)
	19		Woodward Clyde Consultants (1991)
Continuously as biocide	50		Redondo and Lomax (1997)
Shock treatment	500 – 1000		Redondo and Lomax (1997)
Copper sulfate	1.2	<1*	Abart (1993)
	0.1 – 0.5		FilmTec (2000)

*concentration limit for disposal

0.5-1 ppm throughout the whole pretreatment line in RO systems (FilmTec, 2000) and comparable to levels in MSF plants (see Table 3.1). However, disinfecting efficiency of residual chlorine depends on the amount of undissociated species present, which is predominantly HOBr in seawater. As HOBr is less dissociated than HOCl at pH 8 and available at about 85 %, additional acid treatment is not necessary to shift equation 3.5 to the left (Redondo and Lomax, 1997). For this reason, chlorination is very effective in seawater applications.

Dechlorination must be accomplished for RO systems to avoid membrane oxidation, while for MSF plants residual chlorine levels are rejected with the brine. Discharge concentrations are unfortunately rarely published and those cited in Table 3.1 are about one tenth of dosing levels. For dechlorination in RO, sodium bisulfite (SBS, $NaHSO_3$) or sodium metabisulfite (SMBS)[a] may be used. SBS is a strong reducing agent and reacts with hypobromous acid:

$$NaHSO_3 + HOBr \rightarrow NaHSO_4 + HBr \qquad (3.6)$$

A threefold surplus of SBS is practically necessary to remove all free residual chlorine (Redondo and Lomax, 1997), which amounts to dosing of 1.5-3 ppm SBS to protect the RO membrane or 0.6-1.5 ppm if the MSF discharge was treated.

Although chlorine has proven very effective, two major side effects have to be considered, especially in organic-rich water. First, halogenated organic compounds like trihalomethanes (THMs) are by-products, which are problematic from an environmental point of view, and second, chlorine oxidizes organic matter by breaking it down to smaller fragments so that additional nutrients become available. Some desalination plants have faced heavy aftergrowth and attributed it to continuous chlorination, while intermittent chlorination increased plant performance (Redondo and Lomax, 1997; Winters, 1997; Baker and Dudley, 1998; Obaid and Hamida, 1998). Especially RO and MSF systems that operate in water temperatures above 25 °C were affected, e.g. in the Red Sea or Arabian Gulf, (DuPont, 1994; Winters, 1995). As no other reliable pretreatment method has been developed so far, it is uncertain if surface water intakes are possible in hot climate without the current practice of chlorination (Saeed et al., 2000). Therefore, additional shock treatment becomes necessary to re-establish low bacterial numbers from time to time. SBS, that is used as an antioxidant for dechlorination, also reduces bacterial numbers by oxygen depletion and may be used for shock treatment in RO, as no currently implemented membrane is tolerant to high chlorine levels. The recommended dosage is for example 500-1000 ppm for 30 min every 24 hours, but SBS may also be used continuously up to levels of 50 ppm (Redondo and Lomax, 1997). However, it

[a] SMBS ($Na_2S_2O_5$) reacts to SBS in aquatic environments.

is doubtful whether this method is efficient, as some marine bacteria might survive in anaerobic conditions (Baker and Dudley, 1998). Other typical oxidizing chemicals for this purpose are I_2 or H_2O_2 (DuPont, 1994) but again, membranes should not be exposed to high doses. In MSF plants, shock chlorination is common practice and may vary between 1 ppm for one hour a week (Shams El Din and Makkawi, 1998) and 4 ppm for 20 minutes every 8 hours (Abdel-Jawad and Al-Tabtabaei, 1999). Unfortunately, no information is available on residual concentrations of sanitizing chemicals in the brine blow-down but they are presumably higher than during normal operation. If shock treatment does not restore the membrane's performance, off-line cleaning becomes necessary for disinfection and biocides for cleaning are dealt with in section 3.6 on page 40.

Alternative disinfectants to chlorine have been proposed for RO plants, namely chloramine and copper sulfate (Winters, 1995). Chloramine (NH_2Cl) is produced if chlorine is added to ammonium-rich water and some bromochloramine (NHBrCl) will also be present in seawater. As both have germicidal properties, 10-20 ppm NH_4Cl might be added to the raw water before chlorination with 5-10 ppm takes place to increase the amount of halogen amines. The major advantage of chloramine is that it does not degrade high molecular organic compounds to assimilable nutrients, as it is less reactive than chlorine (DuPont, 1994; FilmTec, 2000). Nevertheless, SBS must be added to protect the membrane but dechlorination is not necessary for non-oxidizing biocides like copper sulfate. Improved control of biological growth was reported for a Red Sea plant, which has used copper sulfate since 1986 (Umm Lujj, Saudi Arabia, Saeed et al., 2000). However, no dosage was specified but copper sulfate can be added continuously at levels of 0.1-0.5 ppm as recommended by membrane suppliers. It mostly affects algae but does not remove bacteria sufficiently. Furthermore, copper ions might precipitate as carbonates or hydroxides. Besides these adverse effects on process control, copper salts have detrimental effects on the environment and discharge is restricted in many countries (FilmTec, 2000). The underlying arguments are identical to those discussed in section 4.7 on page 71, which deals with copper contamination caused by corrosion.

TABLE 3.2: Coagulant and polyelectrolyte concentrations in RO plants

RO	Dosing ppm	
Coagulants		
Ferric chloride (40 %)	0.8	Baig and Kutbi (1998)
	25	Woodward Clyde Consultants (1991)
Ferric chloride	3.2	Al-Tisan et al. (1995)
(% not specified)	2.5 – 4.5	Hussain and Ahmed (1998)
	10 – 30	FilmTec (2000); DuPont (1994)
Ferric chloride sulfate	2.51	Ebrahim et al. (1995)
Polyelectrolyte	0.5	Woodward Clyde Consultants (1991)
	0.54	Ebrahim et al. (1995)
	0.2 – 0.4	Baig and Kutbi (1998)
	0.25 – 0.5	Hussain and Ahmed (1998)
	0.4 – 4	DuPont (1994)

3.2 Coagulants

Most particles have a negative surface charge so that they repel each other and do not form larger and faster sinking particles, therefore remaining suspended. To induce particle formation and reduce SDI, coagulants and polyelectrolytes can be added to the feedwater in desalination plants. This pretreatment is common in RO systems while suspended material is less critical in MSF plants, although it may also settle in the heat exchanger tubes, thereby enhancing corrosion and causing their blockade (Al-Awadi, 1994).

Iron(III)-salts, mostly ferric chloride ($FeCl_3$) but also $FeClSO_4$, are used to destabilize the negative surface charge of colloids and to entrap them into the voluminous ferric hydroxide precipitate (Redondo and Lomax, 1997). Aluminum salts ($AlCl_3$) are also effective but coagulation is more difficult to control and therefore could damage the membrane. FeO(OH) and $Al(OH)_3$ precipitation depends on pH which is usually adjusted by adding H_2SO_4. Acid addition is optional for $FeCl_3$ and improves the chemical's performance but for $AlCl_3$, pH control is rather important. Its minimum solubility is in the range of pH 6.5-6.7, being more soluble at higher and lower values so that aluminum ions might easily enter the RO feedwater. This increases the risk for precipitation on the membrane (DuPont, 1994), and therefore $Al(OH)_3$ is generally less recommended by membrane manufacturers although it can be used successfully. For example, a slightly different compound (polyaluminum chloride, $Al_2(OH)_{2.7}Cl_3$, PAC) was found to be more effective than ferric chloride in an RO plant on the German island Helgoland (pers. communication Bomm, 2001).

Polyelectrolytes can be used alone or as a coagulant aid to bridge particles together. Usually, they are soluble organics with high molecular masses, e.g. polyphosphates, polyacrylic acid or polyacrylamides (Falbe and Regitz, 1995). Their functional groups may be anionic or neutral, but cationic polyelectrolytes should be avoided if negatively charged antiscalants are used, as several RO plants have been heavily fouled by precipitation (FilmTec, 2000). The dosage of coagulants is critical to efficient process control. If it is too low, suspended material will not be sufficiently removed from the feed stream but if aluminum and ferric chloride are overdosed, they can foul the membrane. The optimum level depends on plant location and seasonal conditions and is usually between 1-30 ppm for coagulants and 0.2-4 ppm for polyelectrolytes (Table 3.2). Regular monitoring of the silt density index is therefore a sensible way to adapt dosage to the raw water condition.

3.3 Scale control additives

3.3.1 Acid

Raw water for MSF and RO plants is commonly treated with acid as $CaCO_3$ solubility depends on pH. The Stiff & Davis stability index (S&DSI)[b] for seawater can be calculated to determine if $CaCO_3$ tends to dissolve (S&DSI < 0) or precipitate (S&DSI > 0) during the desalination process. Precipitation becomes more likely as the brine concentration increases but can be prevented by acid addition, which keeps the pH of solution below the pH of saturation (FilmTec, 2000). Surplus acid shifts the equilibrium to the right and thereby increases $CaCO_3$ solubility and forces CO_2 out of solution:

$$CaCO_3 + H^+ \rightleftharpoons Ca^{2+} + HCO_3^- \qquad (3.7)$$

$$HCO_3^- + H^+ \rightleftharpoons CO_2 + H_2O \qquad (3.8)$$

Different acids can be used, e.g. hydrochloric acid (HCl), but treatment with sulfuric acid (H_2SO_4) is more common for its low chemical cost (Shams

[b] S&DSI $=$ pH $-$ pH_s, where pH refers to the actual pH and pH_s to the pH at saturation.

El Din and Makkawi, 1998). On the other hand, acid must be added in high concentrations as it reacts stoichiometrically with carbonates. Unwanted side-effects are corrosion of pipes and brine channels or even an increased risk of sulfate scaling. However, the risk of enhanced $CaSO_4$-deposition should be minimal with regard to the natural sulfate concentration of seawater (Spiegler and El-Sayed, 1994), given that overdosing is prevented.

As calcium carbonate is the main scale forming species in seawater RO systems, acid treatment can be sufficient to cope with this problem. Regardless of side-effects, H_2SO_4 is still the most common scale pretreatment in RO plants and can be added exclusively or in support of antiscalants (Al-Shammiri et al., 2000). For example, lowering the pH to 7 allows an increase in conversion to 50 % (DuPont, 1994). Dosage and pH levels vary from plant to plant but concentrations are generally between 30-100 ppm and pH can be down to 6 (cf. Table 3.5 on page 34). As depletion of H_2SO_4 by reaction is intended, low pH values in the brine discharge are not to be expected if acid is not overdosed.

MSF plants face alkaline and sulfate scale formation, of which the latter is predominantly controlled by operational parameters while the first must be chemically inhibited. Thermal decomposition of HCO_3^- starts above temperatures of 50 °C and carbonate ions further degrade above 80-90 °C (Al-Ahmad and Aleem, 1993):

$$2\,HCO_3^- \rightarrow CO_2 + CO_3^{2-} + H_2O \tag{3.9}$$

$$CO_3^{2-} + H_2O \rightarrow CO_2 + 2\,OH^- \tag{3.10}$$

Calcium carbonate is therefore likely to precipitate at relatively low temperatures when carbonate predominates in the above equilibrium (Eq. 3.9), whereas magnesium hydroxide may form with increasing temperature. This is due to a higher availability of hydroxide ions, which are set free by thermal decomposition of CO_3^{2-} as shown above (Eq. 3.10) but also by self-dissociation of water at higher temperature (Spiegler and El-Sayed, 1994). However, Mubarak (1998) states that only negligible amounts of hydroxyl ions are released so that $Mg(OH)_2$ crystallization is slow. Both alkaline scales are again easily controlled by HCO_3^- depletion according to equation 3.8. This method was widely accepted in the 1960-70s (Hamed et al., 1999) but the given disadvantages triggered research for new scale control additives, which are preferred to acid treatment today. In this context, it should be pointed out that acid cleaning is still carried out in MSF plants, and no alternative cleaning procedure has been reported so far.

3.3.2 Antiscalants

Antiscalants were initially developed for use in boiling and cooling water chemistry (Ning, 1999), later implemented as scale control additives in thermal desalination plants and are nowadays commonly approved for RO systems, too.

Polyphosphates were the first antiscalants to be used on a commercial scale, e.g. sodium hexametaphosphate or "Shuwikh Mix" in Kuwait in 1959 (El-Dessouky, 2000). Polyphosphates are like mineral acids replaced by new polymers for the following reasons. Hydrolysis above 60 °C not only lowers their efficiency but also increases the risk of calcium phosphate scaling, which is difficult to remove from heat exchange surfaces (Patel and Finan, 1999). Orthophosphate also serves as a nutrient and could enhance bacterial and algal growth in the plant. In contrast, new polymers like phosphonates and polycarbonic acids are stable and highly effective for scale control in MSF plants with TBTs up to 110 °C (El-Dessouky, 2000; Hamed et al., 1999). However, these new additives are suspected to form calcium deposits, too (Gill, 1999; El-Dahan and Hegazy, 2000). Sodium hexametaphosphate (SHMP) is still in use but more common in RO plants that operate on feedwater temperatures below 45 °C (Al-Shammiri et al., 2000). It is sometimes found to perform better in RO systems than the new antiscalants and this may due to higher brine concentrations or different hydrodynamics in RO (Butt et al., 1997; Hasson et al., 1998).

Regardless of these controversial aspects, polyphosphates, phosphonates and polycarbonic acids all have the same advantage over H_2SO_4 of being highly effective at lower concentrations. Antiscalants do not react in stoichiometric proportions with scales but interact in different physical-chemical ways. Precipitation of salts is considerably delayed as divalent cations (Ca^{2+} and Mg^{2+}) are kept in solution even if solubility products are exceeded (threshold effect)[c]. When initial crystals are formed, further growth is stopped or modified into a fluffy structure as the inhibitor adsorbs

[c] Modern antiscalants allow for a safe operation up to a S&DSI of 1.8 without precipitation (FilmTec, 2000).

to the crystal's surface. It may also impart a like-charge so that particles do not further agglomerate (Patel and Finan, 1999). Thereby, particles are less likely deposited into solid scales and on-line ball cleaning systems in MSF plants continuously remove any sludge that might form. Many inhibitors also sequester other divalent ions like Fe^{2+} or Mg^{4+} by forming soluble complexes (Sheikholeslami, 1999).

However, the underlying mechanisms are still poorly understood, particularly their quantitative aspects (Hasson et al., 1998), and optimum dose levels are usually achieved empirically. A plant specific optimum is desirable as it will minimize scaling and the risk for accidental sludge formation caused by overdosing (Hamed et al., 1999) in addition to cost reduction. The type of treatment, i.e. acid or antiscalant or both combined, the chemical nature of antiscalants and their dose levels not only differ among plants but are also published only fragmentary. Furthermore, numerous products (some are listed in Table 3.3 and 3.4) are commercially available of which chemical composition is often concealed for proprietary rights.

As a general rule, however, required dosage will increases with the plant's TBT and conversion. General dosage recommendations for MSF were corrected downward many times due to better antiscalant performance and awareness of chemical consumption. For example, the Saline Water Conversion Corporation (SWCC) of Saudi Arabia recommended dosing between 4.5-12.5 ppm depending on TBT between 90-110 °C in 1981, which was gradually reduced to 1-2 ppm (Hamed et al., 2000). Other dosage levels for RO and MSF were summarized from a literature review and are provided in Table 3.5 on page 34, which is sorted after chemical groups. The different groups will be further introduced in the following paragraphs, although clear classification is not always possible when substances provide more than one functional group and thereby belong to several categories, e.g. phosphonates with multiple carboxyl groups.

Polyphosphates result from condensation reactions of ortho-phosphoric acid, $(HO)_3PO$, and can be linear or cyclic in structure (Fig. 3.1). Cyclic polyphosphates or metaphosphates have the total formula $(MPO_3)_n$, with M being a monovalent cation like sodium or hydrogen. SHMP, $(NaPO_3)_6$, belongs into this group and the commercial product Calgon is often declared as SHMP but has the structure $(NaPO_3)_{15-20}$ (Falbe and Regitz, 1995). SHMP forms a soluble complex with divalent cations and thereby

TABLE 3.3: Commercial and non-proprietary antiscalants

Polycarboxylic acid (not further specified)	
Belgard EV2030, EV2035 and EV2050	Patel and Finan (1999)
EL-5500	Butt et al. (1997)
Polyacrylic acid (PAA)	
Flocon 100: acrylic acid polymer	Al-Shammiri et al. (2000)
Flocon 200: acrylic acid and organophosphonate copolymer	Luu (1994)
Aqua Feed 600: acrylic acid homopolymer	Luu (1994)
Aqua Feed 1000: acrylic acid based copolymer	Luu (1994)
AA/AMPS: copolymer of acrylic acid and 2-acrylamido-2-methylpropylsulfonic acid	Gill (1999)
Hypersperse-Al: copolymer of acrylic acid and acrylamide	Luu (1994)
DP 3153: copolymer of acrylic acid and sodium phosphonate	Luu (1994)
Darex 41: polymethylacrylic acid	Butt et al. (1997)
Toflo SI-5	Luu (1994)
Pretreat 110	Luu (1994)
Degussa POC 2020	Patel and Finan (1999)
Cynamer p-70	Hasson et al. (1998)
EL-4010	Butt et al. (1997)
Polymaleic acid (PMA)	
Belgard EV: Homopolymer of maleic anhydride	Patel and Finan (1999)
Belgard EV2000: Terpolymer of maleic acid and two aliphatic esters	Shams El Din and Mohammed (1997)
Sokalan PK10i: polymaleic acid based copolymer	Patel and Finan (1999)
Kao KC550: polymaleic acid based copolymer	Patel and Finan (1999)
Organo-Phosphonates	
Bayhibit: 2 phosphono-1,2,4 -butanetri-carboxylic acid $C_7H_{11}O_9P$	Falbe and Regitz (1995)
PBTCA: 2 phosphonobutane-1,2,4-tricarboxylic acid $C_7H_{11}O_9P$	Gill (1999); Butt et al. (1997)
HEDP: 1-Hydroxyethylidene 1-1 diphosphonic acid $C_2H_8O_7P_2$	Gill (1999)
AMP: Amino tri(methylene phosphonic acid)	Gill (1999)
HMDTMP: Hexamethylene diamine tetra(methylene phosphonic acid)	Gill (1999)
PAPEMP: Polyamino polyether methylenephosphonic acid	Gill (1999)
EL-5600: polycarboxylat and phosphonat	Butt et al. (1997)
Albrivap DSB	Patel and Finan (1999)
Aquamax LT-19	Patel and Finan (1999)
Permatreat 191	Al-Shammiri et al. (2000)
Dequest 2010, 2054, 2060, 2066	Luu (1994)

TABLE 3.4: Commercial and non-proprietary antiscalants (continued)

Polyphosphates	
SHMP: Sodium hexametaphosphate $(NaPO_3)_6$	Gill (1999)
Calgon: Sodium hexametaphosphate $(NaPO_3)_{15-20}$	Hasson et al. (1998)
Aqua-Mag: linear $Na_{56}H_3P_{57}O_{147}$	Hasson et al. (1998)
Sequest-All: linear $Na_{56}H_3P_{59}O_{147}$	Luu (1994)
STP: Sodiumtripolyphosphate $Na_5P_3O_{10}$	Falbe and Regitz (1995)
Nalco	Butt et al. (1997)

reduces the hardness of water or complexes heavy metal ions. Linear polyphosphates also have antiscaling properties. Their total formula is $M_{n+2}P_nO_{3n+1}$, but the actual composition can deviate slightly from this, e.g. the antiscalant Aqua-Mag $Na_{56}H_3P_{57}O_{147}$ (Hasson et al., 1998). Another linear but much shorter compound is sodium triphosphate (STP, $Na_5P_3O_{10}$)[d], which is recommended by some membrane manufacturers as a cleaning chemical. STP was a common detergent in washing powder for its water softening quality but has been replaced in many countries by other substances, e.g. by nitrilotriacetic acid (NTA), to reduce phosphate discharge and control eutrophication.

Phosphonates are the salts and esters of phosphonic acid, $HP(O)(OH)_2$, of which two representative substances are shown in Fig. 3.1. These derivates are important in water treatment applications, as they provide one or more phosphonic groups which are linked via C-P bonds to the organic framework. The C-P bonds are thought of being more stable at higher temperatures than the polymeric O-P-O bonding in polyphosphates (Darton, 1997). Despite being effective against scaling by masking Ca^{2+} and Mg^{2+} ions, they are also claimed to inhibit corrosion (Falbe and Regitz, 1995). Similar substances for water softening contain nitrogen, e.g. 1-amino-1,1-diphosphonic acid, which resembles HEDP in structure, and amino tri(methylenephosphonic acid) (AMP)[e].

Polycarboxylic acids are characterized by functional -COOH groups and antiscalants in seawater desalination are usually based on polyacrylic or polymaleic structures (Fig. 3.2). Polyacrylic antiscalants can be derived from homo- or copolymerization of acrylic acid or acrylic acid esters with

[d] syn. sodiumtripolyphosphate and pentasodiumtriphosphate (Falbe and Regitz, 1995)
[e] syn. nitrilo tri(methylenephosphonic acid)

TABLE 3.5: Antiscalant concentrations in RO and MSF plants

MSF	Dosing ppm	At outlet ppm	
Antiscalings N.N.		0.8 – 1	Ahmed et al. (2000) after Khordagui (1997)
	1 – 1.5		Imam et al. (2000)
	1 – 2*		Hamed et al. (2000)
	3 – 7		Al-Ahmad and Aleem (1993)
	2 – 10		Shams El Din and Makkawi (1998)
	4.5 – 12.5**		Hamed et al. (2000)
Polyphosphate	2.2+		Hamed et al. (2000)
	3 – 5++		
	2.5		Abdel-Jawad and Al-Tabtabaei (1999)
Polycarboxylic acid	1.5 – 2		Hamed et al. (1999)
BELGARD EV2000		0.53	Morton et al. (1996)
BELGARD EV2000	1 – 3		Finan et al. (1989)
EV2030, EV2035	1 – 2		FMC (1997)
BELGARD EV2050	1 – 5		
Polymaleic acid	1.8		Hamed et al. (1999)
Polyphosphonate	1		Hamed et al. (2000)
	3		Hamed et al. (1999)
RO			
Sulfuric acid	30		Morton et al. (1996)
	82		Ebrahim et al. (1995)
	6.6 – 97		Al-Shammiri et al. (2000)
	30 – 100		Al-Ahmad and Aleem (1993)
	45 – 70		Al-Shammiri et al. (2000)
	pH 6		Ebrahim et al. (1995)
	pH 6.7		Baig and Kutbi (1998)
	pH 6 – 7		Al-Shammiri et al. (2000)
Polyphosphate			
SHMP	2		Al-Ahmad and Aleem (1993)
	5		FilmTec (2000)
	3 – 5		Al-Shammiri et al. (2000)
	4 – 5		Al-Shammiri et al. (2000)
	5 – 10		Al-Shammiri et al. (2000)
Polyacrylic acid	2.9		Woodward Clyde Consultants (1991)
Flocon 100	1.9		Al-Shammiri et al. (2000)
Phosphonates			
Permatreat 191	1.4		Al-Shammiri et al. (2000)

*recommended by SWCC today **recommended by SWCC in 1981
+optimized level and ++initial plant design (Jubail and Khobar)

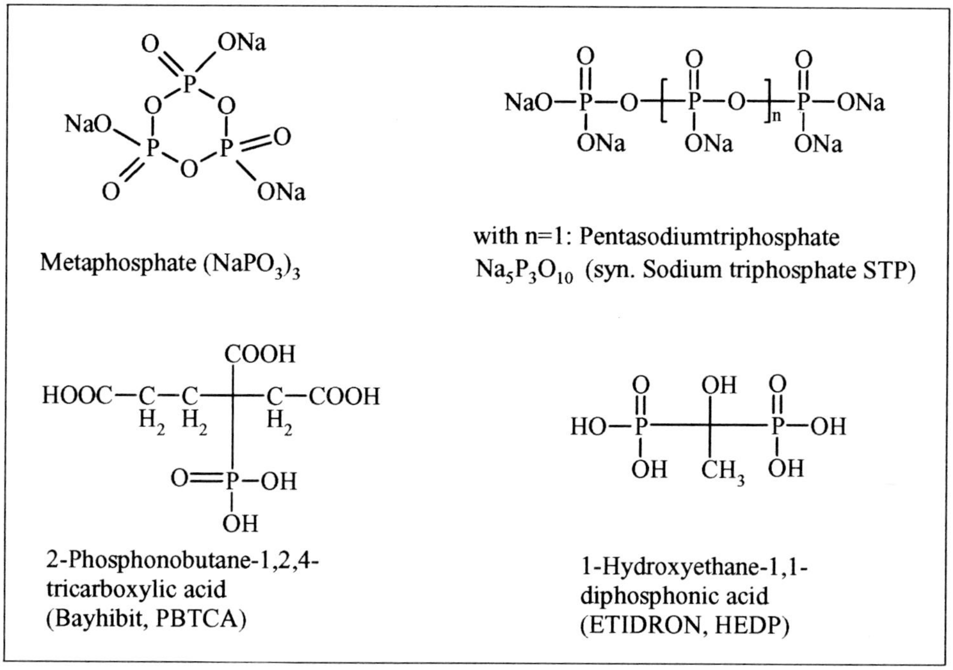

FIGURE 3.1: Polyphosphate (top) and phosphonate antiscalants (bottom). Examples are PBTCA, which is also known as 3-carboxy-3-phosphonoadipic acid or commercially available as Bayhibit (Falbe and Regitz, 1995), and HEDP or Etidron, which is the non-proprietary name of this substance.

other polymers (e.g. phosphonates). These co- or terpolymers are thought of performing better in scale prevention than homopolymers of one single monomer. The ester may be a plain alkyl residual or can contain further amino- or hydroxyl groups. Polymaleic antiscalants are based on maleic acid and corresponding derivates like maleic anhydride and maleates. They can again be homopolymers like Belgard EV (based on maleic anhydride), copolymers with small proportions of other monomers, e.g. acrylates or even terpolymers (e.g. Belgard EV 2000 based on maleic anhydride with two low molecular weight, unsaturated aliphatic esters, Shams El Din and Mohammed, 1997).

As discussed in section 3.2 on coagulants, the same polymers are also widely used as polyelectrolytes. Especially polyacrylic acids and polyphosphates have to be classified as antiscalants on the one hand, which sequester ions and disperse particles, and polyelectrolytes on the other hand, which

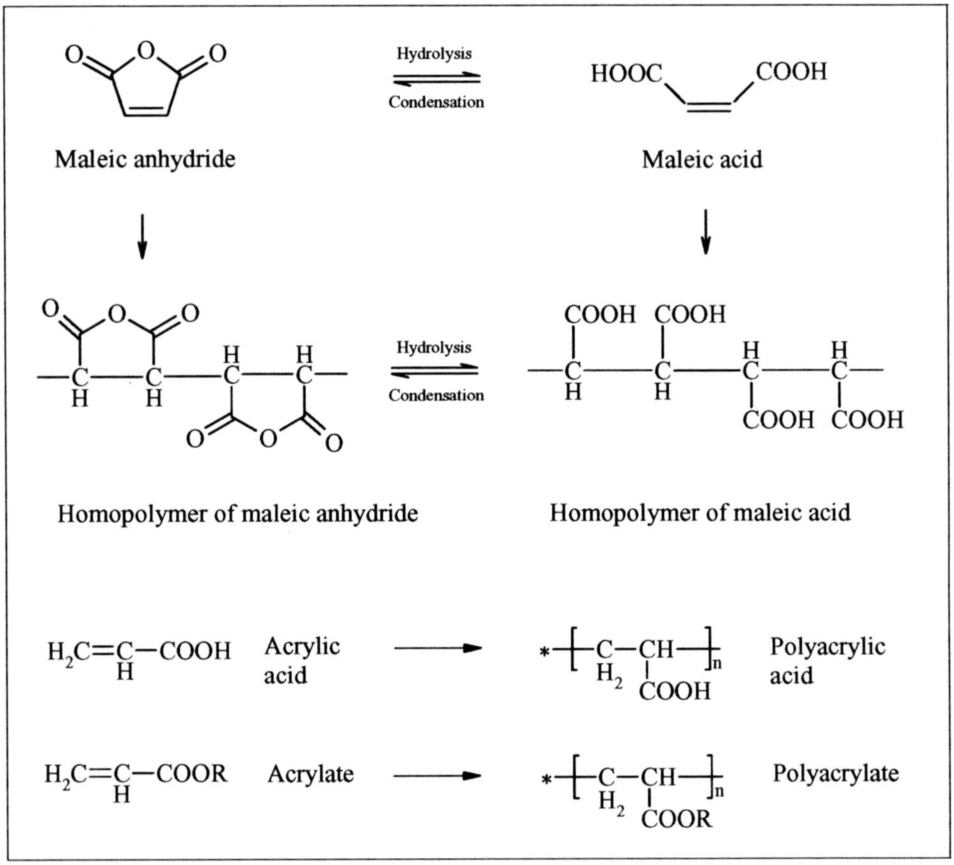

FIGURE 3.2: Polymerization of maleic anhydride and maleic acid (top) and corresponding reactions of acrylic acid and acrylate. It should be noted, that different monomers can be used, so that copolymers and terpolymers are also common additives in water treatment applications as listed in Table 3.3 & 3.4.

enhance agglomeration. Polyelectrolytes are not used in MSF plants but RO systems make use of the coagulating and dispersing properties of these polymers at the same time. Different dosing concentrations may be responsible for the contrasting properties but it has already been said that the underlying mechanisms have not been completely revealed so far. Agglomeration may occur if cationic polyelectrolytes and anionic antiscalants are added to the feedwater (FilmTec, 2000) and in a similar way, polyvalent cations like Fe^{3+}, Al^{3+} or Cu^{2+} and anionic polyelectrolytes may form a precipitate (Falbe and Regitz, 1995). This corresponds to the observation that many anionic antiscalants are not very effective in Ca^{2+} rich waters, as they form insoluble calcium salts that foul RO membranes and MSF heat exchanger surfaces in a similar manner (Gill, 1999). Small amounts of Ca^{2+} or Mg^{2+}, however, gave no apparent reaction with polymaleic acids while large amounts were found to cause turbidity (Shams El Din and Mohammed, 1997). This reveals that polymers can slow down agglomeration but that they also enhance precipitation depending on concentration. Dosage recommendations for polyelectrolytes (e.g. 0.5 ppm) can be ten times lower than those for polymer antiscalants (e.g. 5 ppm) (Table 3.2 and 3.5) and dosage of coagulant (e.g. 5 ppm) can be about a hundred times less than the calcium concentration in seawater (500 ppm). These values result in a coagulant:polyelectrolyte ratio of 10:1 and a calcium:antiscalant ratio of 100:1, which may be the reason for different physical-chemical processes. As mechanisms are not well understood from a quantitative point of view, optimum dosage is usually determined empirically for process control.

This illustrates that each method for scale prevention has to take the overall system into account. H_2SO_4 could increase the risk for corrosion and sulfate scaling, SHMP hydrolyzes easily and thereby provides nutrients and calcium phosphate scales while anionic polymers might precipitate with cationic polyelectrolytes, calcium or dissolved cations, e.g. aluminum or iron to form a gumlike deposit (FilmTec, 2000).

3.4 Antifoaming agents

Antifoaming agents are common in MSF plants but are not used in RO systems, although some chemicals for RO pretreatment like antiscalants provide the same surface active and dispersing properties. Antifoamings are added to disperse the foam-causing organics in the water-air interface, which are present from excretion and degradation processes of phytoplankton in the feed water source (Hoepner, 1999). As the additives are organic substances too, they must be well-chosen so that they do not contribute to foam formation or increase the surface tension of seawater (Falbe and Regitz, 1995). Furthermore, it is required that antifoamings are soluble in water over a wide temperature range: from 30 °C in the low temperature

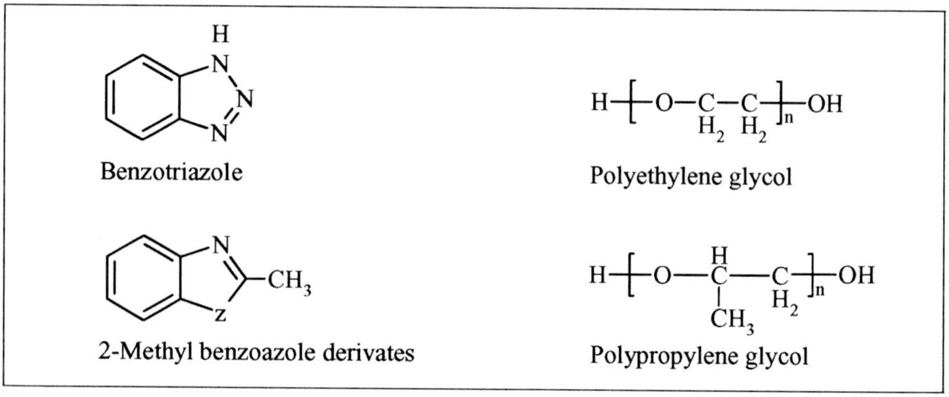

FIGURE 3.3: Corrosion inhibitors (left, z=O: benzoxazole, z=N: benzoimidazole, z=S benzothiazole and z=Se: benzoselenazole) and antifoamings (right).

stages up to 110 °C in the brine heater and they should be compatible with other pretreatment chemicals.

Polyglycol blends are currently in use in some plants of the Saline Water Conversion Corporation (SWCC) of Saudi Arabia. Polyglycol usually refers to polyethylene glycol (Fig. 3.3), which is linear in structure but polypropylene glycol based substances were tested and found to be effective inhibitors, too (Imam et al., 2000). Their chemical properties are determined by their molecular masses in relation to the number of end-standing hydroxide groups. They are generally soluble in organic solvents as well as in water, are very resistant to hydrolysis but sensitive to oxidizing agents like chlorine at high temperatures. Dosage and chemical nature are rarely available for operating plants (see Table 3.6), but addition of less than 0.1 ppm is usually effective. Concentrations in the discharge were found to be half this level, which is predominantly due to mixing of brine with cooling water: While the brine contains residual antifoamings, the cooling water is not treated and thereby reduces the overall discharge concentration.

3.5 Corrosion control

To control corrosion in MSF plants, oxygen levels are brought down to approximately 100 ppb by physical deaeration, which also depletes CO_2 in the feed stream to the heat recovery section. O_2 levels can be further reduced to

TABLE 3.6: Antifoaming concentrations in MSF plants

MSF	Dosing ppm	At outlet ppm	
Antifoaming N.N.	0.1		Pantell (1993)
Polypropylene glycol	0.035 – 0.15*		Imam et al. (2000)
		0.04 – 0.05	Ahmed et al. (2000) after Khordagui (1997)

*test condition of polypropylene glycol-based antifoaming in Al Jubail

about 20 ppb by reaction with oxygen scavengers like SBS (El-Dessouky, 2000), which is transformed into sodium sulfate. SBS is used in RO for dechlorination, and oxygen removal may also be a positive side-effect for protection of stainless steels in these plants. Instead, oxygen depletion is essential for Cu-Ni-heat exchangers in MSF plants to limit corrosion in aerated acidic solution, which is shown in the following reactions corresponding to the electromotive series:

$$\begin{array}{lcll} 0.5\,O_2 + 2\,H_3O^+ + 2\,e^- & \rightleftharpoons & 3\,H_2O & E_0 = +1.23\ V \\ Cu & \rightleftharpoons & Cu^{2+} + 2e^- & E_0 = +0.34\ V \\ Ni & \rightleftharpoons & Ni^{2+} + 2e^- & E_0 = -0.23\ V \end{array}$$

With regard to standard potentials (E_0), the two elements are oxidized simultaneously and Cu^{2+} can also displace nickel from the tubes. By this process, Ni^{2+} concentration may be increased in the brine relative to the alloy's nickel content. This displacement is also known as denickelification (Shams El Din et al., 2000b).

Another preventive measure against corrosion involves addition of chemicals that form a protective film inside the tubes. For example, ferrous sulphate ($FeSO_4$) hydrolyzes and oxidizes to colloidal FeO(OH), which adheres to the surface of condenser tubes. Dosing can be carried out for one hour per day over three months with decreasing levels from 3 to 1 ppm. In this time interval, chlorination and operation of sponge balls should be stopped as both interfere with the formation of the protective layer inside the tubes (Shams El Din and Makkawi, 1998).

Corrosion control is even more important during acid cleaning in MSF plants. In an acidic solution of about pH 2 and temperatures of 60 °C, special corrosion inhibitors are usually required like benzotriazole and its derivates (Al-Mayouf et al. 1998 based on Saline Water Conversion Corporation 1995). These cyclic substances provide multiple bonds and heteroatoms like selenium, sulfur, nitrogen or oxygen with unshared electron pairs (Fig. 3.3 on page 38), which are considered responsible for preventing corrosion. These surplus electrons can interact with metallic surfaces by electrostatic attraction or even electron transfer, so that a stable film is formed that protects the alloy from chemical attack (Al-Mayouf et al., 1998). Dosing recommendations are rare and discharge levels completely lacking, however Andijani et al. (2000) found addition of 25 ppm CP-20 inhibitor effective for protection of titanium tubes (see Table 3.7 on the facing page). This substance provides unshared electron pairs too, and may therefore behave chemically similar to benzotriazole. Inhibitors for water treatment applications are also commercially available, e.g. under the trade name Cobratec CO-99, which contains 100 % solid benzotriazole, or CO-TT-100, which is 100 % solid tolyltriazole (PMC, 2000). The company's general dosage recommendation is 30 ppm for start-up and 3-5 ppm during continuous operation, although this level might be considerably higher for operation at pH 2.

To conclude, the brine discharge will be depleted from oxygen while chemical inhibitors may be present. As corrosion cannot be completely prevented, metals will pass into the brine like Ni^{2+} or molybdate MoO_4^{2-}. Chromates can be either Cr-VI ($Cr_2O_7^{2-}$) in acidic cleaning solutions or Cr-III ($Cr(OH)_6^{3-}$) in alkaline environments, with seawater being slightly alkaline. As these substances are only minor components in metal alloys, their counterparts in the brine will be comparatively rare. It has to be expected that different species of copper and iron are more abundant.

3.6 Cleaning chemicals

Acid is a common additive in both MSF and RO plants for alkaline scale and metal oxide removal, e.g. iron oxide. In MSF plants, the type of acid which is preferred depends on construction material. For copper-based condenser tubes, mineral acids like H_2SO_4 or HCl can be used while titanium tubes have to be treated more cautiously with weaker acids like citric or sulfamic

TABLE 3.7: Corrosion control in MSF and RO plants

MSF	Dosing ppm	
Inhibitor (e.g. CP-20[+])	25	Andijani et al. (2000)
Galvane	N.N.	Shams El Din and Makkawi (1998)
IBIT	N.N.	
Ferrous sulphate	1 – 3	
Cobratec	3 – 5	PMC (2000)
Remaining oxygen	0.02*	Imam et al. (2000)
	0.009 – 0.015**	
	0.045 – 0.047***	
Oxygen scavengers		
Sodium bisulfite SBS	N.N.	
RO		
Zinc orthophosphate	12.8	Woodward Clyde Consultants (1991)

[+] ß-ethyl phenyl ketocyclohexyl-amino hydrochloride *acceptable limit
**after deaeration in Al Jubail II ***after deaeration in Al Jubail I

acid, as titanium adsorbs large quantities of hydrogen (Shams El Din and Makkawi, 1998).

HCl or citric acid are also used to lower the pH to about 2-3 in RO systems for metal oxide and scale removal. On the other hand, pH values up to 12 are effective against biofouling and silt deposits by imposing a like-charge on membrane and foulant, which improves release of particles similarly to conventional washing agents. Other cleaning agents can be broadly subdivided into detergents, complex-forming species and biocides. Recommended combinations of chemicals differ slightly among membrane suppliers and Table 3.8 on page 44 gives an overview of chemicals potentially in use according to the type of fouling. The different chemical categories will be shortly introduced in the following with regard to their purpose in membrane cleaning solutions. As these substances also become waste products and may be discharged to the sea, their environmental impacts are discussed in section 4.8 on page 79.

For example, sodium dodecylsulfate (NaDSS, Fig. 3.4) is preferred for inorganic colloids and organic matter by Filmtec while Dupont recommends a similar detergent, sodium dodecylbenzene sulfonate (NaDBS), in combination with sodium perborate ($NaBO_3$) and sodium hypochlorite (NaOCl) for the same purpose. Organo-sulfates and -sulfonates are surface active

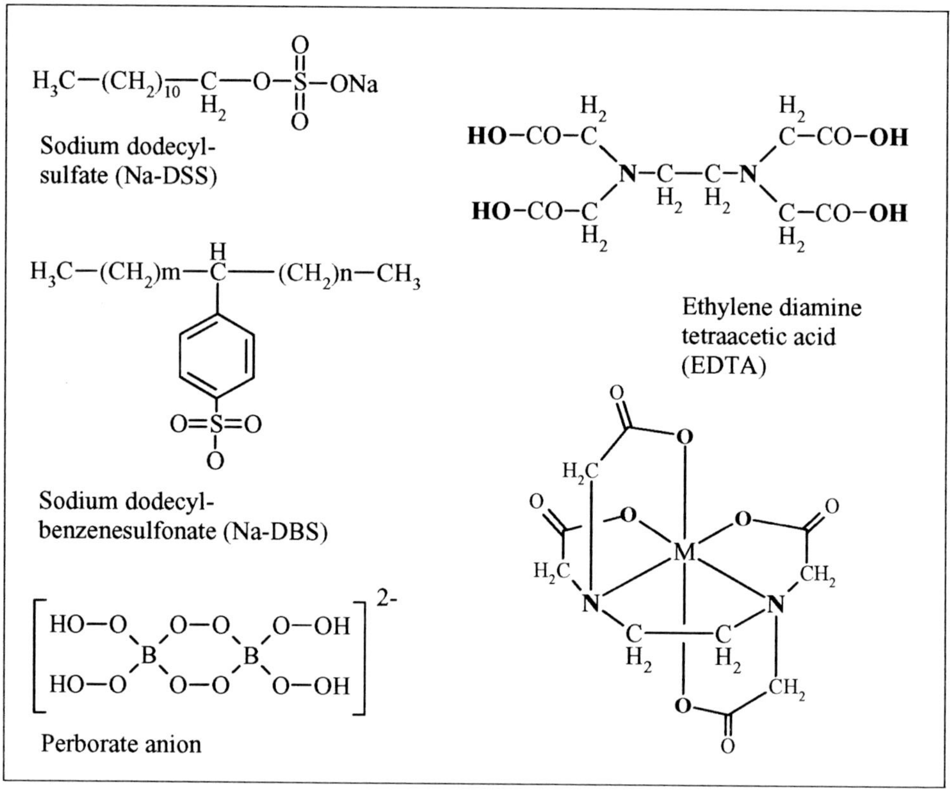

FIGURE 3.4: Chemicals for membrane cleaning: detergents (Na-DSS and Na-DBS), complexing agent EDTA and perborate as a conventional bleaching agent.

agents by providing a lipophilic as well as an anionic, hydrophylic residue. With these different properties, dirt particles are easily removed from the membrane and dispersed in solution. They are common ingredients in washing powder in addition to sodium perborate[f], which is used for its oxidizing (bleaching) quality. In water, it degrades rapidly to sodium borate and H_2O_2 and the latter further breaks down to oxygen and water. By oxidation, sodium perborate improves removal of heavy organic fouling but has to be handled carefully to avoid damage of polyamide membranes (Trisep, 1999).

As an alternative for biological deposits, sodium ethylenediamine tetraacetic acid (Na-EDTA) may be used alone (FilmTec, 2000) or in support

[f] syn. with sodium perborate tetrahydrate ($NaBO_3 \cdot 4\ H_2O$)

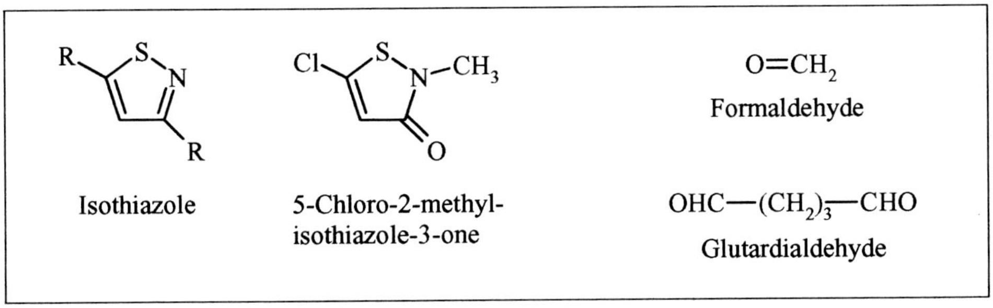

FIGURE 3.5: Biocides for membrane cleaning

of sodium triphosphate and trisodium phosphate (STP: $Na_5P_3O_{10}$, TSP: $Na_3PO_4 \cdot 12\ H_2O$, DuPont, 1994). Similar to citric acid, EDTA chelates metal ions like calcium and magnesium and thereby masks them in solution. This process does not only reduce the hardness of water and decalcifies metallic surfaces, it also removes metal ions that bridge dirt particles together, which are then more easily broken apart. As phosphates like STP or TSP have the same complexing properties, they were once essential in washing powder but have been replaced to reduce eutrophication. They additionally provide soil dispersing properties (Falbe and Regitz, 1995), which increase their efficiency and make them popular cleaning agents for RO membranes.

Biocides for pretreatment like sodium hypochlorite are also used during cleaning at pH 12 and are enhanced by detergents which help to remove deposits. As this method partly oxidizes polyamide membranes, posttreatment with polyvinyl methyl ether and tannic acid becomes necessary to restore the membrane (DuPont, 1994). Other non-oxidizing biocides for membrane disinfection or storage are formaldehyde and glutaraldehyde (Fig. 3.5), which can be used for long-term membrane storage as well. Isothiazole derivates like 5-chloro-2-methyl-isothiazole-3-one are commercially available under the trademark Kathon, which is an assortment of fungicides, bactericides and algicides. It is commonly used for biofilm destruction in cooling water (Falbe and Regitz, 1995) and might therefore be used in MSF plants as well. However, implementation could only be confirmed for disinfection and storage in RO systems (Abart, 1993; Hydranautics, 2000; Trisep, 1999).

TABLE 3.8: Membrane cleaning chemicals

Low pH (2–3)	Filmtec[1]	Dupont[2]	Hydra-nautics[3]	Toray[4]	Ionics[5]	Trisep[6]
$CaCO_3$						
HCl (0.2–0.5 %)	+	+			+	
$C_6H_8O_7$ (1–2 %)	+[7]	+	+	+	+	org. acid
H_3PO_4 (0.5 %)	+	+				
$Na_2S_2O_4$ (1 %)	+	+				
NH_2SO_3H (0.2 %)		+				
Metal oxides						
$C_6H_8O_7$ (1–2 %)		+	+	+	+	org. acid
H_3PO_4 (0.5 %)	+	+				
$Na_2S_2O_4$ (1 %)	+	+				
NH_2SO_3H (0.2–1 %)	+	+				
SHMP (1 %)		+				
High pH (11–12)						
Inorg. colloids (silt)						
HCl (0.5 %)		+			+	
$C_6H_8O_7$ (1–2 %)		+			+	
NaOH (0.1 %)	+				+	+
NaDSS (0.025 %)	+					
NaDBS (0.25 %)		+				
$NaBO_3$ (0.3 %)		+				
NaOCl (0.04 %)		+				
STP (2 %)		+	+[7]		+	
SHMP (1 %)		+				
Na-EDTA (0.1–1 %)			+[7]		+	+
Biofouling						
NaOH (0.1 %)	+	+		+	+	+
NaDSS (0.025 %)	+			+		
NaDBS (0.25 %)		+				+
$NaBO_3$ (0.3 %)		+	+			+
NaOCl (0.04 %)		+				
STP (1–2 %)		+	+[7]	+	+	
TSP (1 %)		+		+	+	
Na-EDTA (1 %)	+	+	+[7]		+	+
Disinfection/storage						
Formaldehyde (0.1–1 %)		+	+	+		
Glutaraldehyde (0.1–1 %)		+	+			
Isothiazolin / Kathon (1–1.5 %, 15–25 ppm)			+			+
SBS (1 %)		+	+	+	+	
H_2O_2 (0.2 %)			+	+		
Propylene glycol (20 %)					+	
Glycerin (20 %)					+	

[1] FilmTec (2000), [2] DuPont (1994), [3] Hydranautics (2000), [4] Toray (1989), [5] Woodward Clyde Consultants (1991), [6] Trisep (1999), [7] Ning (1999), concentrations in first column based on [1–3]; $C_6H_8O_7$ citric acid, $Na_2S_2O_4$ sodium hydrosulfite, NH_2SO_3H sulfamic acid, SHMP sodium hexametaphosphate, NaDSS sodium dodecylsulfate, NaDBS sodium dodecylbenzene sulfonate, $NaBO_3$ sodium perborate, NaOCl sodium hypochlorite, STP sodium triphosphate ($Na_5P_3O_{10}$), NaEDTA sodium ethylenediamine tetraacetic acid, TSP trisodium phosphate (Na_3PO_4), SBS sodium bisulfite.

4

Impacts of Desalination Plants

In the previous chapters, process-related chemicals were identified. Their residuals and reaction by-products are present in the discharge in addition to the concentrated seawater constituents. The chemically and physically modified seawater has become a waste product and with regard to the immense volumes, return to the sea is usually the only practicable method for disposal. The effluent properties must therefore be within certain limits so that the receiving water can meet water quality standards. These are prerequisite for protecting aquatic life and sustainable use of the marine resources, including drinking water production. It is the interest of this chapter to address possible impacts of effluent disposal in addition to regulatory means which intend to minimize the adverse effects on the marine environment.

The extent of regulations in different countries reflects the awareness for environmental protection as well as its feasibility with regard to a country's economic situation and need for potable water, as far as desalination is concerned. The USA for example, have the tenth largest desalination industry regarding worldwide capacities (Wangnick, 1999), and resulting emissions are regulated by well-developed environmental standards. Consequently, it is not surprising that these standards may be consulted for impact assessment of discharges outside the USA, including those from desalination (e.g. in Mannaa, 1994) and similar legislation could become implemented for a better protection of marine resources from land-based activities in other parts of the world. A short description of some regulatory means will therefore be given in the following section as necessary for the purpose of this chapter, which is the assessment of single parameters of desalination plant discharge with regard to water quality criteria, mixing zone definitions and toxicological information.

4.1 Regulations for desalination plant discharge

Usually, new plants have to be integrated into an overall management program of the coastal zone and require some kind of environmental impact assessment, which also includes aspects of waste disposal. The corresponding laws are, for example, the US Coastal Zone Management Act and National Environmental Policy Act. However, discharge from point-sources into water bodies is also subject to regulation under a separate permit (NPDES[a]) in compliance with the Clean Water Act, and all desalination processes have to seek permission due to their discharge of brine and pretreatment chemicals (EPA, 2001).

Theses effluent components can basically be subdivided into two pollutant categories, although classification may be indistinct. Salinity, total suspended solids, temperature, pH and biodegradable, oxygen demanding organics represent normal seawater properties and are therefore considered 'conventional', although levels in the desalination discharge usually deviate from natural background levels. Subsequently, a change has to be expected in the area surrounding the outfall, usually an elevation of the specific parameter, which may provide unfavourable environmental conditions for aquatic life. In contrast, some 'toxic pollutants' like copper are essential at low concentration. It is the environmental concentration of each pollutant which is of crucial importance for aquatic life. Although pollution can strictly be defined as a man-made increase in concentration above natural background levels, it is usually the impact which shall be prevented. Consequently, pollution is commonly considered as the introduction of such amounts of material that will have a deleterious effect on aquatic life (GESAMP, 1986). Both definitions are not easily applicable for the following reasons: (1) natural backgrounds must be distinguishable from man-made contamination and (2) cause-effect relationships between environmental levels and certain effects are not always easily determined.

[a] National Pollutant Discharge Elimination System permit

Regulations usually intend to minimize pollution by adopting water quality standards that are considered safe for marine life, as they are based on a comprehensive review of current knowledge. These standards, however, may not be obligatory in a restricted area around the outfall, which can be established on a case-by-case basis. This "allocated impact zone" (according to US EPA, 2001) shall allow for sufficient dilution and degradation of pollutants, so that safety criteria apply at the edge of the mixing zone and provide protection for the remaining water body. If some effluent components cause an excursion above water quality criteria, a limitation has to be included in the discharge permit (WQBEL[b]). Therefore, the discharge permit, which focuses on end-of-pipe properties, has to suit the requirements of the mixing zone, and both must be designed to allow for an adequate protection of the water body.

Regulatory mixing zones are often applied to industrial discharges (Jirka et al., 1996) and can be considered for desalination plants, too, as the US Environmental Protection Agency (EPA) initiated classification of membrane concentrate as an industrial wastewater (Kimes, 1995). This term is often considered misleading for the argument, that the waste contains only high concentrations of what is already in the water (Ahmadi, 1994), but this assumption disregards residual pretreatment chemicals in the blow-down. Although these substances contribute only minor amounts to the overall discharge, some may be hazardous to the marine environment. Both categories of pollutants must be addressed, and disposal of 'toxic' substances is usually more strictly handled than effluent limits for conventional parameters.

4.1.1 Water quality criteria and mixing zone definitions for toxic pollutants

As toxicity depends on concentration on the one hand and time of exposure on the other, regulations have to implement time-dependent criteria. In the concept of US WQC, this is taken into account by maintaining one criterion for long-term (indefinite) and another for brief exposure, which are the CCC and CMC[c]. They estimate the highest concentration of a

[b] Water Quality Based Effluent Limitation

[c] Criterion Continuous Concentration and Criterion Maximum Concentration

substance in surface water that will not produce any unacceptable effects to the aquatic community.

If toxic pollutants are discharged, the continuous criterion (CCC) may be exceeded within the mixing zone that has been established for conventional pollutants, but acutely toxic concentrations are generally prohibited as well as deposition of unacceptable loads of effluent constituents. Lethal conditions are consequently avoided, if the end-of-pipe concentration is below the specified criterion for brief exposure (CMC). As an alternative for meeting this limit, a toxic dilution zone may be granted where the CMC may be exceeded. Although this zone seems to be incompatible with the prevention of acutely toxic levels, the CMC is strictly defined as an allowable concentration for short exposure. Strict regulations for the toxic mixing zone shall therefore enhance dilution, which counteracts long exposure times. In addition, the CMC must be met within 10 % of the distance that is allowed for an excursion of the CCC (or 1 % of a circular mixing area) and five times the local water depth, which shall prevent outfall location in shallow coastal water (Jirka et al. (1996) after EPA, 1991). Mixing zones are therefore made to measure the local conditions including hydrodynamics and topography while areas of importance like nursery grounds or shellfish harvesting areas can be excluded. Numeric values may limit this zone, e.g. a maximum of about 125,000 m^2 for coastal waters or four times this area for the open ocean may be granted (e.g. in Florida, Kimes, 1995), which corresponds to a radius of about 200-400 m and 20-40 m respectively. If no binding spatial restriction exists, it must be assessed if the actual size of a specific plume allows for a reasonable protection of the water body.

4.1.2 A procedure for impact assessment of hazardous substances

Mixing zone definitions and water quality criteria can be utilized for an impact assessment of chemical residuals which might be hazardous to the marine environment. As water quality criteria (WQC) apply to the receiving water body, effluent concentrations must also be projected to the area around the outlet, i.e. likely environmental concentrations must be predicted in the mixing zone. Subsequently, both levels can be compared and if environmental concentrations are found to be higher, discharge must be restricted for protection of the water body (Fig. 4.1).

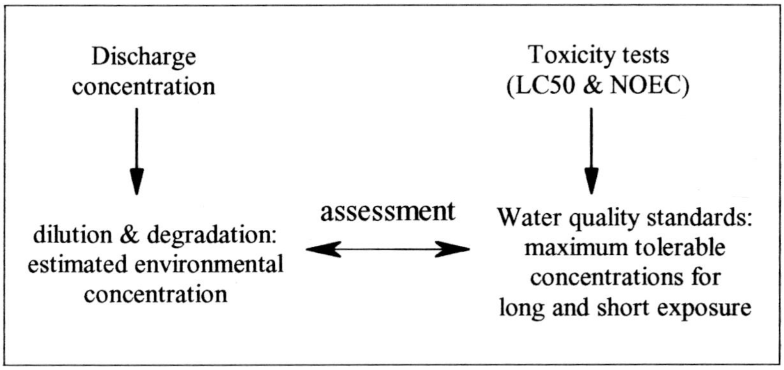

FIGURE 4.1: Impact assessment of hazardous chemicals.

For some pollutants, seawater quality criteria are recommended by the US EPA, which will be consulted in the following sections. These maximum allowable limits are derived from toxicological data on at least eight families out of a wide spectrum of aquatic species, so that the criteria can be applied to different water bodies and will protect the majority of organisms (EPA, 1998). If no criterion has been established so far, the risk of deleterious effects on the environment will be assessed by consulting toxicity test results, which will inevitably produce a wide range of concentrations depending on test organisms and observed effects. The corresponding parameters are usually the LC_{50} as a measure for acute toxicity while the No Observed Effect Concentrations (NOEC) refers to the level where no chronic effects can be observed[d].

It is far more difficult to estimate the environmental concentrations adequately than to obtain WQC or toxicological information, although these are also rare for some substances. Discharge permits clearly specify limits and operators of desalination plants are obliged to periodically monitor a wide spectrum of parameters (Mickley, 1995), but the results are usually not published, at least not for all parameters. However, a mosaic of process-related chemicals was derived from a literature review and presented in the previous chapter with obtainable discharge concentrations. If these sub-

[d] The LC_{50} is the concentration of a chemical in water, which is lethal to 50 % of all test animals within a given time span of commonly 24, 72 or 96 h. In contrast to acute tests, chronic toxicity tests last considerably longer depending on the reproductive cycle of the test organisms, i.e. several days to months, as observed symptoms are for example growth inhibition and reproduction rate (Fent, 1998).

stances are hazardous to aquatic life will be assessed by comparing the effluent concentrations with water quality criteria and toxicity test results. Cleaning chemicals (section 4.8), which are not used on a daily basis, will be discussed in the same way as pretreatment chemicals (4.3-4.7) and furthermore, characteristics of the brine like salinity, temperature, pH and dissolved oxygen will be included (4.2).

Although salinity and temperature are dealt with in almost every study related to environmental impacts of desalination plants (e.g. Ahmed et al., 2000; Hoepner, 1999; Abdel-Jawad and Al-Tabtabaei, 1999; Morton et al., 1996; Mickley, 1995; Shams El Din et al., 1994; Del Bene et al., 1994; Mannaa, 1994; Altayaran and Madany, 1992), these parameters will not be omitted as they are main effluent characteristics. More important however, dilution of other effluent components cannot be discussed without these two properties, as they set the basic conditions for spreading and mixing of the plume through density differences with the receiving water body. This discussion can only be of general nature, as the plume is also influenced by local conditions like bottom topography, shoreline and water currents. The practice of discharge is also important: Immense volumes of cooling water are usually blended with the concentrate, so that a dilution factor has to be estimated which often depends on the installed electricity generating capacity in dual purpose plants. Therefore, brine properties are modified before they enter the sea and effluent volumes are considerably increased. For example, a numeric maximum like 125,000 m^2 is not applicable to large desalination plants, which return about 3-4 million m^3 per day to the sea. The mixing zone had to be 24-32 m deep without taking power plant cooling water and additional seawater for mixing into account. Without exact knowledge of the specific discharge situation, the area of initial dilution can hardly be determined. However, it can still be concluded that the mixing zone will extend over large areas, if outlet sites are situated in shallower coastal water.

To be able to discuss impacts as they might occur in the receiving water body, dilution factors can be derived from salinity levels. A reduction of brine concentration is a primary concern of permits which regulate desalination plant effluents, and adequate mixing with cooling water usually decreases salinity levels to about ambient. The same dilution factor will be assumed for residual chemicals in the brine in the following sections.

However, some wastes have to be disposed of periodically like filter deposits and cleaning solutions. It is unknown, if these materials are dumped all at once or if they are blended with the effluent stream, which requires extra storage facilities. Both alternatives will be considered, and a theoretical mixing zone of 125,000 m^2 will be assumed in the first case. The estimated concentrations can then be compared to toxicological information while the calculated volumes will give an impression of how large the impact area could be. As effluent regulations often restrict concentrations on the one hand and flow volumes on the other, total loads per day or per cleaning event are also of interest. Mass limits may especially be preferred for bioaccumulative or poorly degradable substances in addition to those materials that come up intermittently like cleaning wastes.

Most substances will be discussed in the light of actual capacities of existing plants. The Al Jubail RO plant in Saudi Arabia was chosen in addition to the Doha West MSF plant in Kuwait, as the two are among the largest plants in the world and situated in shallow coastal environment, so that the investigation of marine impacts is of special concern. Furthermore, obtainable information was more detailed than for many other plants and locations, e.g. regarding cleaning of Dupont permeators (DuPont, 1994) as well as seawater properties or already existing impact-related studies. This approach will allow for a comparison between chemicals as well as different processes, and the identification of substances which are of special concern for the marine environment (cf. section 4.9).

4.2 Brine characteristics

4.2.1 Salinity

Effluent salinity and temperature depend on the process of desalination. As thermal plants have lower recovery rates and consume water for cooling, the seawater flow rate has to be 3-4 times higher than the feed to RO membranes for the same amount of product water extraction (Morton et al., 1996). Consequently, the discharge of MSF plants is less saline than the brine from the RO process. For example, Fig. 4.2 illustrates that an MSF plant with a conversion of 10 % would reject 90 % of the intake water with an increased salinity of 50, if the feedwater had a salinity of 45. This high value is typical for the coastal waters of Kuwait (Abdel-Jawad and

Al-Tabtabaei, 1999), where the Doha West plant is located, and representative for shallow seas in warm climate, which are characterized by high evaporation and little freshwater input during the summer months. In contrast, an RO plant with a conversion ratio of 32 % would reject less water of higher salinity, i.e. 66 under the same assumptions. For comparison, the coastal waters near Al Jubail have an average salinity of 43.8 (Baig and Kutbi, 1998) so that typical brine concentrations will be about 64 g/l total dissolved solids (Table 4.2).

The immense difference in effluent salinity is predominantly due to mixing with once through cooling water from the heat rejection section in MSF plants (cf. 2.2 on page 16 & Fig. 2.2). Although the brine blow-down from the last MSF stage has a salinity of almost 70, which is the operational upper limit to prevent scale formation (El-Dessouky, 2000), the concentration of salts is diluted with a threefold amount of cooling water of ambient salinity, so that a more moderate value of about 50 can be discharged. The given examples are in line with Mickley (1995), who states that MSF brine is usually within 15 % above the ambient seawater salinity.

Further dilution of the effluent occurs in the receiving water body, or within the co-sited outfalls of dual purpose power and desalination plants. Large MSF plants are usually close to electricity generating stations to satisfy their high energy demand[e]. Therefore, cooling water of ambient salinity from the power plant further reduces the amount of total dissolved solids before the combined effluent enters the sea. In the case of Doha West, the brine is more than ten times diluted to a salinity of 46.9, which is only slightly above ambient (Fig. 4.2, Table 4.1 on page 80). RO plants can be located near power stations, too, and shared intake and outfall channels will not only reduce operation costs but will also mitigate the impacts of concentrate discharge, as shown for the Tampa Bay RO plant in Florida with a capacity of 95,000 m^3/day (Graves et al., 2001). However, RO plants of moderate or low capacity will predominantly be single-purpose plants. Although they produce less concentrate, adequate mixing must be ensured. The seawater for the Al Jubail RO plant is also drawn from existing intake

[e]MSF plants are three times as energy-intensive as RO plants at the same amount of product water extraction (Al-Mutaz, 1996) due to pumping of larger seawater volumes and the distillation process itself. Primary energy is used more efficiently if the steam from the turbine, which produces electricity, is utilized as thermal energy for desalination (Darwish, 2000).

structures of adjacent MSF dual purpose plants (Baig and Kutbi, 1998), but it is not known if outfalls are also co-sited. Irrespective of where dilution takes place, a tenfold amount of seawater will yield an overall salinity of slightly above ambient.

4.2.2 Temperature

In contrast to the RO process, MSF effluent temperature will generally be increased. Reported temperature differences varied between 5-6 °C, 6-9 °C and 10-15 °C above the naturally occurring seawater temperature in summer and winter (Abdel-Jawad and Al-Tabtabaei (1999); Mickley (1995); Altayaran and Madany (1992) respectively). As the reject cooling water from the last MSF stages as well as the power plant cooling water are similarly warmed, mixing will not lower the overall effluent temperature. On the contrary, about ten times the brine volume originates from the cooling process.

4.2.3 Mixing zones, density differences and biological effects

As shown above, blending of different waste streams can adjust concentrations of total dissolved solids to more ambient levels, but it must be assumed that not all plants provide the same mixing conditions. For example, a high salinity up to 55 was reported in the receiving water close to the outlet of an MSF plant, which is 10 g/l above the salinity of adjacent control sites, but values were back to normal within 160 m distance (Altayaran and Madany, 1992). A larger distance was mentioned by Shams El Din et al. (1994), as elevated salinity and temperature decreased within 700 m distance from the outlet. These findings indicate, that a considerable degree of mixing can still occur in coastal waters. If a distinct mixing zone exists, its extent will depend on several parameters like plant size or local water depth. In the given examples, the first distance was observed near the Sitra desalination plant in Bahrain, which has about one third of the capacity of Doha West, while the latter refers to the Umm Al Nar plant

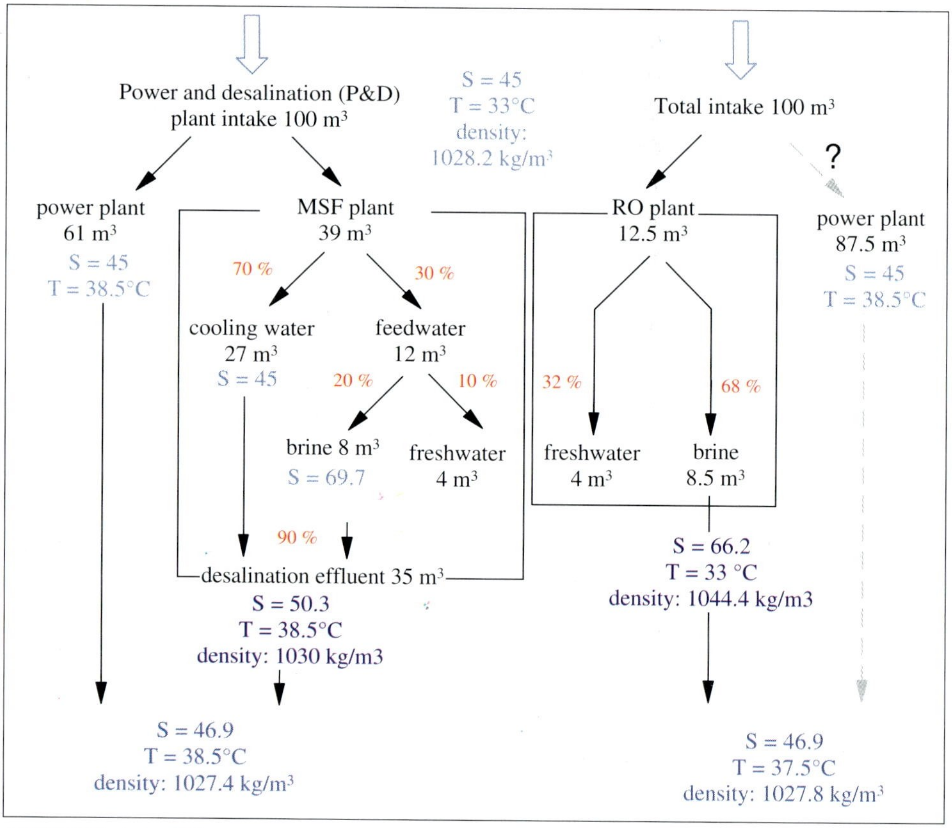

FIGURE 4.2: Water intake, conversion ratios and effluent characteristics of typical MSF (left) and RO plant design (right). While dilution with cooling water decreases brine salinity, all waste streams (except RO brine) show increased temperatures (+ 5.5 °C, Abdel-Jawad and Al-Tabtabaei, 1999). The overall effect is a decrease in density (cf. Fig. 4.3).

in the UAE, which is comparable to Doha[f]. As all plants are situated in shallow coastal areas, higher discharge rates can account for larger plumes.

Furthermore, plume spreading depends on density differences to the receiving water. MSF plumes can be neutrally buoyant as observed by Morton et al. (1996), if salinity and temperature effects on density are balanced, but are more likely to float on the sea surface if temperatures are

[f] Sitra: 126,000 m^3 freshwater per day, 125 MW (Altayaran and Madany, 1992)
Doha West: 436,465 m^3, 2400 MW (Abdel-Jawad and Al-Tabtabaei, 1999)
Umm Al Nar: 384,000 m^3, 1130 MW (Shams El Din et al., 1994)

too high to be compensated for. Altayaran and Madany (1992) reported such a surface plume, which must be due to the high effluent temperature of 10-15 °C above ambient, but even 5 °C will be sufficient as shown in Fig. 4.3. The designed mixing distance was also doubled as vertical circulation was restricted. Strong horizontal spreading is a common feature of positively buoyant plumes (Jirka et al., 1996), so that high volumes of heated desalination and power plant cooling water will probably cover a large surface area.

Effects on drifting organisms must be considered, which pass through the plume during the mixing process and are not able to avoid it. The elevated temperature of MSF discharge and power plant cooling water will boost biological activity during winter, whereas summer values could be less favorable for survival when metabolic rates increase (Morton et al., 1996). For example, average seawater temperatures in the Arabian Gulf range from 12.5 °C in February to 32 °C in August (Subba Rao and Al-Yamani, 2000) and Fig. 4.3 shows monthly averages in Kuwait Bay, where seasonal variations are similar. The corresponding rise in effluent temperature of 5 °C (Abdel-Jawad and Al-Tabtabaei, 1999) will result in temperatures between 19 and 38 °C in the mixing zone, which is still below the critical limit of 39 °C. This level is assumed to cause an abrupt decline in biological activity in Gulf water (Shams El Din et al., 1994). Although Abdel-Jawad and Al-Tabtabaei (1999) conclude that temporary lethal effects are outweighed by positive impacts on marine life by the current practice, thermal discharge still influences the natural seasonal change in temperature considerably. In the same way as increased nutrient levels are positive for marine organisms at a first glance but can cause serious problems, thermal discharge is generally regarded as pollution. In addition, the thermal increase can be as high as 10-15 °C according to Altayaran and Madany (1992), so that more critical situations cannot be excluded elsewhere. Therefore, means which enhance dilution are of great importance for MSF plant discharge and minimize impacts by restricting the spatial extension of the plume.

In contrast to MSF plumes, highly saline RO discharges will sink to the sea floor where mixing could be restricted by bottom topography. Proper dilution, e.g. by different outlets and high discharge velocities, is therefore of paramount importance for RO process water if mixing with other waste water is not possible. If the effluent has a higher density than the coastal water, sensitive benthic communitics could be affected. Many subtidal organisms exposed to the discharge are also found in the intertidal and are therefore adapted to similar variations. However, they might not withstand high salinities permanently so that a continuous discharge might alter the benthic community. For example, bivalves can close their shells during harsh conditions but finally reopen them for respiration and feeding. Iso et al. (1993) investigated the survivorship of soft clams and found no lethal effect at a salinity of 50 but more than 50 % of the test organisms died in a salinity of 60 within 48 hours. The same study indicates that salinities below 45-50 had no effect on several other test organisms, however, salinity above 43 was assumed to be critical to coral species like *Porites lutea*, which is also prominent in the Arabian Gulf and Red Sea (Sheppard et al., 1992). These findings stress the need for proper dilution prior to discharge, but to which extent ?

It is difficult to determine the effect of increased salinity on marine organisms which are more or less able to cope with saline water, and monitoring studies in the discharge area of desalination plants in the Arabian Gulf are not known. However, observations from naturally hypersaline environments in the Gulf indicate that levels exceeding 45 alter the benthic community considerably, e.g. in the Abu Dhabi barrier islands complex (UAE, Evans et al., 1973) or the Gulf of Salwah, where many species of corals, molluscs and echinoderms were found to be absent (between Qatar and Saudi Arabia, Clarke and Keij, 1973). Lower species diversity and abundance were also reported by Coles and McCain (1990) for salinities above 45, and the authors stressed the importance of salinity as an environmental factor, even for the "relatively hardy marine species" in the Arabian Gulf.

With regard to these results, it can be concluded that brine discharge from the last MSF stage or from an RO plant will have acute effects on marine organisms if it is not diluted. A salinity of 55-50, as observed near Sitra and usually achieved by dilution with reject cooling water, will increase the environmental stress for marine organisms, although it might not be acutely toxic. This level would also not comply with EPA regulations, as the increase in salinity must be restricted to 10 % of the background salinity (Graves et al., 2001). Therefore, mixing with power plant effluent is strongly recommended if possible, which will reduce effluent salinity to

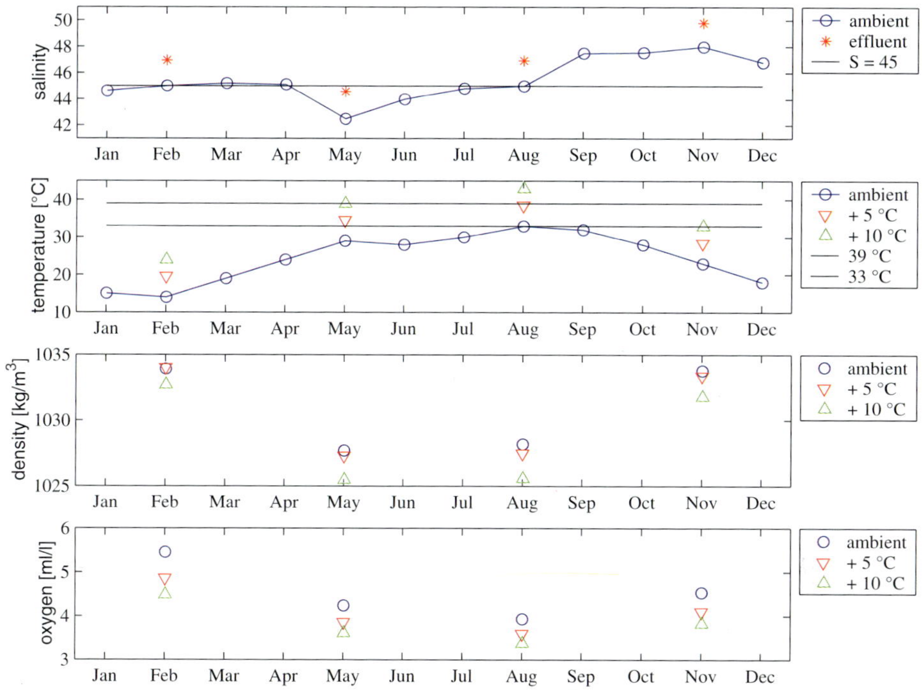

FIGURE 4.3: Monthly average seawater salinity and temperature values in Kuwait (graphs in upper two diagramms, according to data from Abdel-Jawad and Al-Tabtabaei, 1999). Based on these, effluent properties were calculated (single symbols): effluent salinity takes a varying conversion ratio into account (between 12-9.4 % depending on ambient salinity of 42-48, El-Dessouky 2000) but flow rates were assumed to be constant; calculation of density and oxygen solubility (saturation) according to Morgan (1998, 1994). Horizontal lines show threshold values for temperature and salinity: while 33-34 °C is optimal for biological activity in the Gulf, 39 °C or more will cause an abrupt decline (Shams El Din et al., 1994); an increase in salinity above 45 throughout most of the year might alter the benthic community. Salinity and temperature values result in a positively buoyant plume as shown in the density plot.

slightly above ambient (+ 1-2 g/l, Fig. 4.2). It is often assumed that variations of ± 1 can be tolerated (Del Bene et al., 1994; Mickley, 1995) but an upper limit must be taken into account. Although a salinity of 46 or more can still be considered a normal concentration in many areas of the Gulf during summer, it should be investigated if most organisms can cope with levels above the threshold of 45 throughout most of the year. Although dilution with power plant cooling water minimizes the impact of salinity, it increases the amount of thermal input to the coastal water and thereby enhances biological activity throughout most of the year. Critical situations may occur from August to October, when ambient seawater temperature and salinity are already high. Effluent discharge will enhance this seasonal trend, especially as water and power consumption (for air-conditioning) are highest during the summer months (Abdel-Jawad and Al-Tabtabaei, 1999). A high density difference to the surrounding water impairs mixing of the plume, which extends over larger distances. As a consequence, its chemical residuals are poorly diluted and will increase the stress for marine life.

4.2.4 Dissolved oxygen

With increasing temperature and salinity, oxygen becomes less soluble in seawater as shown in Fig. 4.3, however, the major decline is due to deaeration in MSF plants and dosing of oxygen scavengers like SBS for dechlorination in RO. Although levels in MSF are usually reduced to 0.02 ppm, Altayaran and Madany (1992) observed oxygen levels of about 5 mg/l at the outlet, which is only slightly lower than ambient and may be due to mixing with cooling water. For RO brine, the lack of dissolved oxygen is considered toxic by US authorities and aeration is required prior to oceanic discharge (Kimes, 1995). However, no further remarks on this topic or measurements of oxygen levels adjacent to RO plant outlets could be found.

4.2.5 pH

Significantly lower pH values may occur during acid cleaning rather than during normal operation of MSF and RO plants, as acid reacts stoichiometrically with scales. Altayaran and Madany (1992) reported no significant pH alterations in the outfall site of a power and desalination plant. In contrast, Mabrook (1994) attributed a change in density and diversity of marine organism to desalination discharges and stressed the need for pH adjustment prior to blow-down, as the pH of the last stage was 5.8 ± 0.25 compared to 8.3 ± 0.3 in coastal waters. Periodic cleaning will further deteriorate seawater quality in certain intervals but no information on the practice of discharge or direct observations are available. Assuming that 5000-7000 m^3 seawater are acidified to pH 2 to clean one distiller (Shams El Din et al., 1994), about 20,000 m^3 of seawater will be necessary to restore a natural pH of 8 if the waste is returned to the sea[g]. For cleaning a large plant like Doha West with 16 units and a capacity of 436,000 m^3/day, a total seawater volume of 320,000 m^3 would be required for subsequent neutralization, which equals an area of 125,000 m^2 of about 2.5 m depth. For comparison, Kuwait Bay is on average about 4.5 m deep (Ali and Riley, 1986).

In RO plants, a pH range of 2.3-4 is achieved by adding hydrochloric-, phosphoric-, sulfamic or citric acid to the cleaning solution (DuPont, 1994). For example, a total water volume of 584 m^3 is acidified for a large plant the size of Al Jubail (cf. section 2.1.4 on page 15). Depending on pH, between 23 m^3 and 1200 m^3 seawater are needed for neutralization, which is less than 0.5 % of the seawater volume necessary for Doha West. As Al Jubail has about 20 % of Doha's capacity, the acidic discharge is relatively lower for RO with regard to plant size. This is due to generally lower cleaning volumes and higher pH values.

Even though environmental effects of cleaning waste disposal have not been adequately investigated so far, an impact seems likely in the time until the discharge is neutralized (Shams El Din et al., 1994). For fish toxicity, the amount of acid is more important than the type of acid. Sulfuric acid, hydrochloric acid and phosphoric acid at a pH of 3-3.5 caused 50 % mortality of bluegill within 96 h and swimming ability was impaired at sublethal concentrations (HSDB 2001a). Comparable LC_{50} values were reported for starfish (HCl at pH 2-2.5/48 h) and salt water prawn (H_2SO_4 at pH 3-3.3/48 h) (HSDB 2001b & c). While fish may be able to avoid the discharge plume, starfish and other less mobile organisms will be heavily affected by the acidic blow-down. It may be assumed that not all MSF distillers are cleaned simultaneously but on the other hand, acidic waste will be released every three weeks on average based on a distiller cleaning frequency of once per year. Therefore, neutralization or at least blending with the brine during normal operation should be considered.

[g] Seawater has a concentration of 150 mg/l HCO_3^- or $2.5 \cdot 10^{-3}$mole/l in the coastal waters of the UAE (Shams El Din et al., 1994), correspondingly 20,000 m^3 seawater provide 50,000 mole HCO_3^- for reaction with 50,000 mole H^+ in 5000 m^3 discharge with a concentration of 0.01 mole/l (pH 2).

4.3 Biocides

4.3.1 Chlorine

Residual chlorine, or more precisely hypobromite and -chlorite, is degraded by reaction with organic or inorganic compounds and self-decomposition to chloride ions and oxygen. While the first process gives rise to harmful, halogenated compounds which are discussed in the next section, the second process limits the impact of chlorine on the aquatic environment.

Chlorine is present in the discharge of MSF brine and cooling water, and effluent levels are periodically increased by shock treatment in the plant. After an initial rapid decrease of this load in seawater, degradation occurs at a reduced rate. For example, Shams El Din et al. (2000a) observed a high decomposition up to 90 % in direct sunlight and warm seawater within 45 minutes. Assuming that about one tenth of the initial chlorine dosage is discharged, about 200-500 μg/l residual chlorine could be present in the waste water at blowdown (cf. Table 3.1 on page 24). This load will be further diminished by degradation and an additional loss of 90 % would yield a concentration of 20-50 μg/l, not accounting for dilution.

Although residual chlorine is generally recognized as problematic, few studies investigated chlorine concentrations adjacent to desalination plant outfalls so far. Ali and Riley (1986) reported levels of 30-100 μg/l and Abdel-Jawad and Al-Tabtabaei (1999) estimated a level of 50 μg/l in 1 km distance from the outlet, both referring to the coastal waters of Kuwait. Although these concentrations seem to be very low, it cannot be concluded that the "impact on the marine environment is insignificant" (Abdel-Jawad and Al-Tabtabaei, 1999), which would justify the current practice of chlorination. On the contrary, chlorine is highly toxic even at low levels and this conclusion can be verified by several toxicological results as shown in Fig. 4.4. In agreement with these findings, chlorine is classified as a pollutant by the US EPA, for which a low short-term allowable maximum concentration (CMC) of 13 μg/l as well as 7.5 μg/l for long-term exposure (CCC) in saltwater was specified (EPA, 1998). With regard to WQC, it must be concluded that a discharge concentration between 200-500 μg/l does not comply with the acute criterion in the outfall channel, so that levels will be further decreased in a 'toxic dilution zone' (cf. section 4.1.1). However, this toxic zone covers large coastal areas as estimated environmental levels

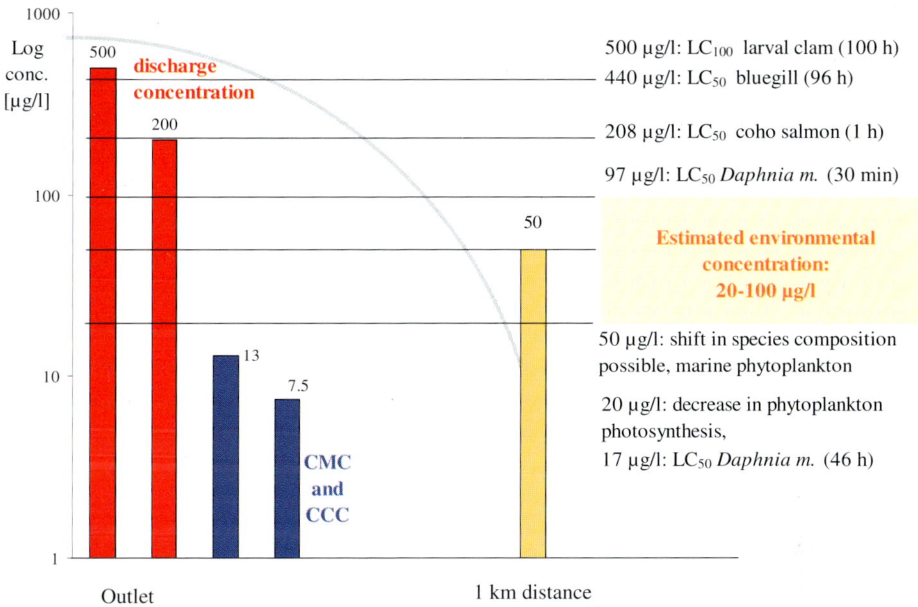

FIGURE 4.4: Comparison of chlorine concentrations in the discharge, water quality criteria and toxicological data: marine phytoplankton show different sensitivity towards chlorine and a shift in species composition may occur at levels of 50-150 μg/l (HSDB 2001d). Growth inhibition of phytoplankton was also reported in aged water (35 days), although no remains of the initial chlorine level of 10 mg/l could be detected (HSDB 2001e), and photosynthesis may be reduced by 80 % in residual chlorine levels of 20-40 μg/l in coastal discharge sites of power plants (Kennish (1997) after Eppley et al., 1976). Zooplankton tests showed that LC_{50} values of *Daphnia magna* ranged from 97 μg/l/30 min to 17 μg/l/46 h. Lethal concentrations for marine fish may be higher, e.g. LC_{50} values of 208 μg/l of coho salmon (1 hour only) and LC_{50} of *Lepomis macrochirus* (bluegill) at 440 μg/l/96 hr (HSDB 2001d & f), but levels as low as 10 μg/l were also reported (Kennish (1997) after Mattice and Zittel, 1976).

(e.g. 50 μg/l) still exceed the acute criterion by a factor of 4 in 1 km distance. The aspect of chlorine dispersion due to plume spreading is resumed in section 5.3 in the chapter on the Arabian Gulf.

As major impacts on marine life have to be expected by chlorination, mitigation methods should be considered and the current practice re-evaluated. For example, chlorine is like other biocides not permitted for discharge in California and must be treated prior to blow-down (Pantell, 1993). Dechlorination is achieved in RO plants by SBS to protect polyamide membranes. Nevertheless, a limit for residual chlorine in the RO effluent is proposed and shall not exceed 10 μg/l (EPA, 1999), so that the acute criterion will be met prior to blow-down. The same practice could be established in MSF plants as well. However, overdosing should be prevented as SBS is slightly hazardous and may cause oxygen depletion, which is also toxic for marine life.

Shams El Din and Mohammed (1998) discussed alternative neutralization chemicals, for example SO_2 reacts with hypochlorite and finally yields HCl and H_2SO_4. Although it should be added with care to prevent unnecessary pH reduction, low dosage is probably of no concern and acids will be neutralized by seawater alkalinity as discussed in section 4.2.5 on pH. Another reaction with H_2O_2 forms water, oxygen and chloride and overdosing should again be avoided as H_2O_2 is also a biocide, yet weaker than chlorine. The major advantage of H_2O_2 over SO_2 is its lower chemical cost, even though SO_2 is a by-product from fossil fuel combustion in power plants. In the worst case, neutralization of 0.5 mg/l residual chlorine in 9 m^3 discharge with the more costly SO_2 method would cost about 10 US cents based on the estimate given by Shams El Din and Mohammed (1998). This extra expense in addition to slightly higher operating costs for pumps etc. might amount to about 10 % of the product costs for 1 m^3 potable water[h]. For example, drinking water is produced at 155 cents/m^3 in a Saudi Arabian MSF plant (Al-Mutaz, 1996). As costs will probably be lower, it must be concluded that impact mitigation by dechlorination is economically and technically feasible. A trade-off between operating costs and environmental protection is necessary. This must be seen in the context that water is highly subsidized in Saudi Arabia: cost recovery is minimal, as tariffs are only collected if water consumption exceeds 0.5 m^3/day and billing is restricted to 65 % of the total water supplied (MEDRC, 2001).

4.3.2 Halogenated hydrocarbons

By-products like halogenated hydrocarbons from chlorination are usually discussed for MSF discharge only and little attention has been given to RO plants, in which chlorination is a common pretreatment method, too. Residual chlorine levels are neutralized for protection of polyamide mem-

[h] 9 m^3 desalination discharge have to be treated for 1 m^3 product.

branes, but not for cellulose acetate membranes, and secondary reactions are likely even before dechlorination has been completed. The focus on MSF plants is founded on their higher chlorine consumption, as they require more feed water, and the continued reaction in the coastal water body. Halogenated hydrocarbons arise either from reaction with precursors of natural origin, which will predominantly produce trihalomethanes (THMs) or with petroleum components, for example giving rise to chlorophenols or chlorobenzenes (Saeed et al., 1999). However, the number of by-products can hardly be determined due to many possible reactions of hypochlorite and -bromite with organic seawater constituents, and the already complex chemistry of chlorine in seawater is deteriorated by high oil pollution in the Arabian Gulf (Hoepner, 1999).

A recent study investigated the levels of halogenated hydrocarbons in coastal waters of Kuwait and the results indicate that levels and composition depend on desalination plant location. THMs accounted for more than half of all halogenated compounds in the discharge site of Doha, and about 90 % of the THMs were observed to be bromoform ($CHBr_3$) while dibromochloromethane, bromodichloromethane and chloroform contributed only minor amounts. THM levels varied between 0.001-2.7 μg/l in Kuwait Bay and were considerably higher close to the outlet, ranging between 0.1-9.5 μg/l. As this exceeded the intake concentrations (ND-0.2 μg/l), the contaminated plume was probably carried away from the plant (Saeed et al., 1999). However, the enclosed nature of Kuwait Bay generally restricts circulation and dispersion, so that THMs could accumulate and consequently levels should be lower elsewhere. For comparison, the Al-Zor MSF plant discharges less water and is situated on a more exposed shoreline further south in Kuwait, which could account for the tenfold difference in average THM concentrations.

In a previous study, considerably higher THM levels up to 83 μg/l were reported for the outfall site of Shuwaikh in Kuwait Bay (Fig. A.7), which decreased to less than 1 μg/l within 2-4 km distance (Ali and Riley, 1986). As the high level of THMs correlated with a high environmental concentration of 500 μg/l residual chlorine and occurred during flood tide, a build-up of contaminants seems possible. It should be noted, that 500 μg residual chlorine per liter is also the upper limit in the Doha West discharge, but environmental concentrations are expected to be lower due to degradation and dilution following discharge. Lower environmental chlorine levels could explain why THM levels were found to be lower near Doha.

Although bromoform and other THMs can be produced naturally by marine algae at significant quantities (for example a production of about 0.08 μg/h and gram algae wet weight was observed for coralline algae, HSDB 2001g), it is evident that desalination plants further increase natural background levels. The main processes that reduce THM concentrations in the environment are dilution and evaporation, as biodegradation, bioaccumulation and adsorption to suspended particles and sediment are of minor importance (HSDB 2001h-i). Therefore, it has to be investigated if elevated THM levels pose a risk to marine life in the mixing zone before ambient levels are achieved. As no water quality criteria were obtainable for THMs, acute and chronic concentrations will be consulted and compared to the observed environmental levels.

With regard to relatively high LC_{50} values of several test organisms, e.g. 29,300 mg bromoform/l/96h in static bioassay for *Lepomis macrochirus* (HSDB 2001j), acutely toxic situations will not occur in the discharge site of desalination plants. Different species, however, can be more sensitive and especially oysters seem to be easily affected by bromoform. Oyster larvae of *Crassostrea virginica* had an LC_{50} of only 1 mg/l/48 hr (static bioassay, HSDB 2001k) while chronic effects of adult oysters were observed at even lower concentration of 25 μg/l: The animals' respiration rate was increased and feeding rate and size of gonads were reduced (Scott et al., 1982). Even if higher sensitivity is taken into account, chronic effects seem to be very unlikely in the area of highest environmental concentrations. As sufficient evidence exists that bromoform is an animal carcinogen, which is also assumed for other THMs based on their structural similarity (HSDB 2001k), further attention might be given to chronic effects. However, the acutely toxic effects of residual chlorine will predominate over impacts from elevated THMs, for example the LC_{50} of chlorine is by several orders of magnitude lower than for bromoform as observed on *Lepomis macrochirus*. If residual chlorine is neutralized in the discharge, major impacts will be eliminated and side-effects reduced.

4.3.3 Chloramine

Another chlorination by-product in MSF discharge is chloramine, which is formed in ammonium-rich water e.g. close to sewage outfalls, or chloramines may be deliberately added as biocide in RO plants. As neither effluent nor environmental concentrations were obtainable, chloramine will be assessed relatively to chlorine toxicity. For example, *Daphnia magna* did not survive in 0,007 mg/l residual chlorine with monochloramine as the dominant species, which is less than the already low LC_{50} of 0.017 mg/l/46 h for chlorine (HSDB 2001m). However, exposure time was a week instead of two days and LC_{50} values usually decrease with time of exposure. Both chloramine and chlorine might therefore have comparable toxicity. Similarly, the LC_{50} of bluegill was three times higher than the respective value for chlorine within a quarter of the chlorine exposure time (1.5 mg/l/24 h compared to 0.44 mg/l/96, HSDB 2001n). Irrespective of bioassay conditions, the point is that chloramine could be similarly harmful in the marine environment and will not be an alternative to minimize desalination plant impact. Consequently, it is only another option for process control which may be used in organic-rich water to reduce the formation of assimilable nutrients, as it is less reactive than chlorine. Dechloramination is nevertheless required to protect polyamide membranes (DuPont, 1994), which will also protect marine life from the RO effluent.

4.4 Coagulants

4.4.1 Ferric and aluminum chloride

Coagulant wastes from RO desalination plants accumulate during operation in media filters and are disposed of intermittently by backwashing. Ferric chloride is commonly used in RO, and is also a conventional additive for sewage or surface water treatment to remove phosphate and thereby reduce eutrophication (Falbe and Regitz, 1995). As it is not toxic to seawater organism (Pantell, 1993), increased turbidity is of main concern and could adversely affect marine life when the waste is discharged back to sea. Accordingly, no numeric water quality criterion exists, but the wastewater should be free from substances that settle in objectionable amounts or produce objectionable color and turbidity (EPA, 2001).

Ferric chloride reacts to ferric hydroxide, which forms a voluminous precipitate. According to Baig and Kutbi (1998), dosing of $FeCl_3$ (40 %) is done in the RO plant in Al Jubail in very low concentrations of 0.8 mg/l, so that a total amount of 92 kg is added every day[i]. As backwashing lasts about 40 minutes in Al Jubail, it may be assumed that filters are cleaned only once a day and deposits could be discharged all at once. If ferric chloride reacts stoichiometrically to FeO(OH), a total of 50 kg will be held back by filters per day[j] (Table 4.2).

This amount is undoubtedly minimal compared to natural sources of turbidity, e.g. riverine transport by the Shatt Al-Arab[k], but this material is mostly shed before it enters the Gulf (Purser and Seibold, 1973). Thereby, large intertidal areas have evolved in the northern Gulf and the influence on suspended material ceases well before the more southern areas of the Gulf. For example, the seawater close to Al Jubail was of relatively good quality with a background concentration of less than 1 mg/l total suspended solids. With an intake of 288,000 m^3 and assuming that all suspended material is withhold by the media filters, 288 kg solid waste originates from the seawater itself. Additional coagulants increase this amount by 17 %. As coagulant dosage aims at removing suspended solids, it is probably proportional to the natural water turbidity and can be as high as 30 mg/l (Table 3.2 on page 27). The relative increase in natural turbidity is consequently not of major concern if the discharge is diluted properly. However, the intermittent disposal of daily coagulant loads in addition to the accumulated natural material from the filters could be problematic for marine life. For comparison, dilution of 338 kg in an impact area of 125,000 m^2 yields a concentration of 2.7 mg/l in a water column of 1 m, thereby increasing the ambient concentration considerably (Table 4.2). FeO(OH) is reddish-brown in color and might cause a discoloration of seawater as well as a decrease in light penetration. Therefore, a high turbidity near the outlets might impair photosynthesis or increased sedimentation could bury sessile organisms, and especially corals are sensitive to high levels of sus-

[i] Assuming a daily seawater intake of 288,000 m^3 which is effectively dosed at 0.32 mg/l $FeCl_3$ (100 %).

[j] With a molecular mass $M_{FeCl_3} = 162$ g/mole, 92 kg corresponds to 567 mole, so that 567 mole of FeO(OH) with $M_{FeO(OH)} = 89$ g/mole correspond to 50 kg.

[k] The Tigris and Euphrates, which are two of the three rivers that form the Shatt-Al Arab, carry alone about $62.4 \cdot 10^6$ t sediment per year (Reynolds, 1993).

pended material. Impacts can be minimized if the deposits are discharged continuously into well-mixed water or by removal as solid sludge.

Aluminum chloride can be used alternatively as coagulant but is not recommended by membrane manufacturers as it has a higher potential for fouling the membrane. This is in line with lacking dosing information from a literature review. For completeness, it should be mentioned that toxicity depends on Al^{3+} concentration, which is increased at pH < 5 (Fent, 1998). Aluminum toxicity is discussed in the context of acidification of freshwater systems but this is probably not of concern in seawater. Although aluminum chloride hydrolyzes to aluminum hydroxide and hydrochloric acid, the acid will be neutralized by seawater alkalinity. Thereby, toxic effects can probably be neglected in the marine environment as insoluble aluminum hydroxide is not of major concern.

4.4.2 Polyelectrolytes

Auxiliary chemicals for coagulation like polyacrylamide, polyacrylate or polyphosphate are not toxic at levels applied in RO plants, as supported by toxicological information in section 4.5.1 & 4.5.2 on scale control additives. As overdosing should be avoided for membrane protection, most polymers will be attached to coagulant flocs and free residual concentrations are probably low. Therefore, this extra material will further increase turbidity when the backwashing waste is discharged to the sea and sedimentation is enhanced due to the formation of larger particles. Baig and Kutbi (1998) specified that coagulation aid is added at an average level of 0.3 mg/l in Al Jubail, which amounts to 86.4 kg/day (Table 4.2). Overall, 424 kg solid waste have to be disposed of: 68 % are of natural origin and 32 % man-made by adding coagulants and polyelectrolytes. In the presence of chlorine, polyelectrolytes could be broken down and by this process, nutrients or halogenated organics will be released. Otherwise, suspended material will sink to the sea floor, where polymers are finally biologically degraded.

4.5 Scale control additives

4.5.1 Polyphosphates

Sodium hexametaphosphate (SHMP, $(NaPO_3)_6$), which is still used in RO plants as antiscalant, depolymerizes in water to sodium orthophosphate and sodium trimetaphosphate ($(NaPO_3)_3$, HSDB 2001o), and the latter will be further degraded in the marine environment. Thereby, SHMP provides an essential macro-nutrient to marine primary producers. According to Al-Shammiri et al. (2000), antiscalant is used in the Al Jubail RO plant but the type of antiscalant and concentration are not specified. Based on a low dosing of 2 ppm SHMP, Al Jubail might consume 576 kg SHMP per day (Table 4.2). In addition, phosphates are used in membrane cleaning solutions (Table 3.8) and for Al Jubail, a 1 % solution of SHMP amounts to 5,840 kg per cleaning event. The same amount will be necessary if sodium triphosphate (1 %) or trisodium phosphate (1 %) are used, and these surplus nutrients could enter the coastal water from time to time if the cleaning waste is discharged without treatment. However, no information on actual discharge from desalination plants is available. As enhanced primary production and algae mat formation near MSF plant outfalls could clearly be attributed to phosphate dosing (Shams El Din et al., 1994; Abdel-Jawad and Al-Tabtabaei, 1999), the same impact must be considered for RO plants in which polyphosphate antiscalants are still used.

4.5.2 Polymer antiscalants

Use of polymer antiscalants was primarily initiated for their higher efficiency at high TBTs, but Shams El Din et al. (1994) reported also that the problem of algal growth totally disappeared at the outlet of MSF plants with this improved treatment. According to material safety data sheet information provided by manufacturers, polymer antiscalants have relatively high LC_{50} values, i.e. can be classified as comparatively safe for aquatic life. It seems that toxicity depends on pH, as Sokalan and neutralized Belgard EV products had considerably higher LC_{50} values than polymaleic acid of pH 2 and the same observation was made for Flocon 100, a liquid product based on polyacrylic acid (Fig. 4.5). A maximum dosage of 10 ppm will therefore be neutralized by seawater alkalinity immediately. The amount of residual monomers might have an effect on toxicity, too. For example, about 10 % of the solid content of Belgard EV can be inactive monomers (Shams El Din and Mohammed, 1997). Phosphonates like HEDP (etidronic acid) are commercially available as antiscalants, e.g. as 10-20 % solution with a pH of 2.8 under the trade name Hypersperse MDC220 (BetzDearborn, 2000). According to toxicological information provided by the manufacturer (Fig. 4.5), organophosphonates can be classified as non-hazardous similar to other polymer antiscalants. It can be concluded that discharge levels of desalination plants are far below concentrations that could have any significant toxic or chronic effects on marine organisms.

Nevertheless, the environmental fate of polymer antiscalants is of interest and can be assessed by consulting bioaccumulation and -degradation test results, as conducted for some commercial products (Wilcock and Finan, 1994; FMC, 1994; BetzDearborn, 2000). For example, biodegradation of polymaleic acid is slow with 18 % in 35 days (Zahn-Wellens test[l]), whereas polyacrylic acid was degraded almost three times faster (52 % in 35 days), both yielding carbon dioxide and water. Polymer-based antiscalants are unlikely to accumulate in marine organisms, as the tendency for accumulation increases with lipophilic properties of the chemical. Evidence for this behaviour is given by high water-solubility and direct test results that indicate low bioaccumulation factors and octanol/water partition coefficients[m] for phosphonate-, polymaleic and polyacrylic acid-based products.

When polymer antiscalants are dispersed in the environment, they can be compared to naturally occurring humic substances. Both have high molecular masses[n] and functional carboxyl groups (Fig. 3.2), so that chemical properties might be similar, too, even though humic substances provide aromatic ring structures which do not exist in polymer antiscalants. Furthermore, humic substances are resistant to microbial degradation and have a half lifetime of roughly 1 month (GKSS, 1986), which is similar

[l] OECD-Guideline 302 B: a chemical is inoculated with activated sludge, usually for 28 days. Degradation is measured by DOC or COD analysis and given in % of the initial concentration.

[m] $K_{OW} = \frac{\text{concentration in 1-octanol}}{\text{concentration in water}}$ is a measure of the lipophilic property of a substance. 1-octanol is used as reference as its chain length and number of hydroxy groups are characteristic for other natural organics, e.g. cell membranes (Fent, 1998).

[n] For example, polyacrylic acid has a molecular mass of 2000-300,000 g/mole compared to humic acids of 2000-500,000 g/mole (Falbe and Regitz, 1995).

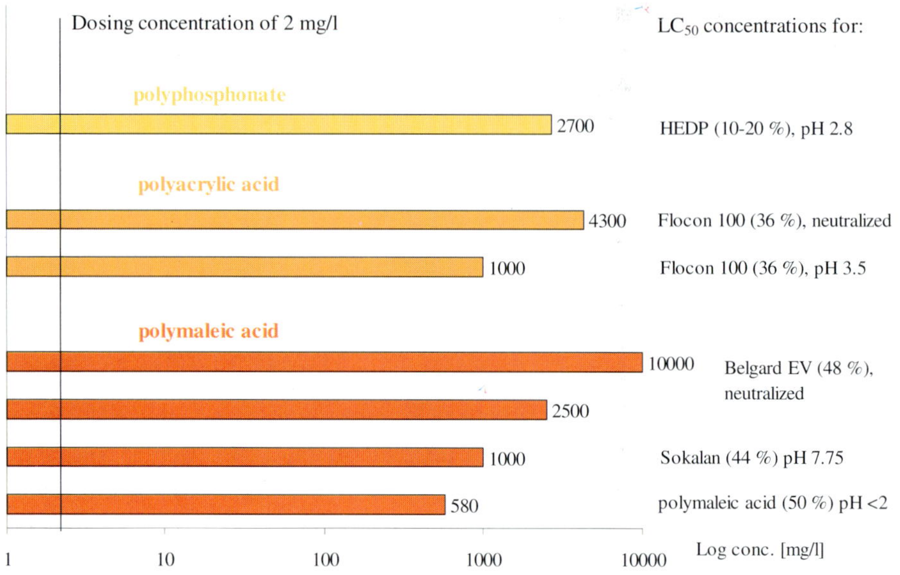

FIGURE 4.5: Polymer antiscalant concentrations of toxicological importance: Polymaleic acid (50 %, pH <2, LC_{50} of 580 mg/l, tested on brown shrimp for 96 h, Redox-Chemicals, 2001). Commercial antiscalants based on polymaleic acid (e.g. Sokalan, 44 %) or polymaleic anhydride (e.g. Belgard EV, 48 %) are slightly less concentrated and have higher LC_{50} values depending on pH: *Brachydanio rerio* 1000 mg Sokalan/l/96 h (BASF, 1994); bluegill 2500 mg neutralized Belgard EV/l/96 h and brown shrimp 10,000 mg neutralized Belgard EV/l/96h (Wilcock and Finan, 1994). The LC_{50} for neutralized Flocon 100 was 4300 mg/l (Wilcock and Finan, 1994) and considerably higher than 1000 mg/l (FMC, 1994), which was established for the acidic solution (both tested on bluegill for 96 h). Phosphonates like HEDP are also comparatively non-hazardous with an LC_{50} of 2700 mg/l/48h for *Daphnia magna*, and a no observed effect concentration of 1540 mg/l/48 h (BetzDearborn, 2000).

to the observed degradability of antiscalants. As antiscalants are added as dispersing and complexing agents, these properties will not cease after discharge. For comparison, heavy metal ions are also complexed by humic acids, which thereby prevent metal precipitation and increase mobility in the environment. Consequently, polymers might influence dissolved metal concentrations in the coastal environment in the same way as naturally occurring substances. To conclude, polymer antiscalants are non-hazardous and comparable to naturally occurring humic material in terms of chemical properties and environmental fate. They may be preferred over polyphosphates from an environmental point of view, as they help to control nutrient input to coastal waters, but their influence on metal solubility needs further investigation. In this context, high discharge loads are of concern and this aspect is resumed in section 5.4 in the next chapter.

4.6 Antifoaming agents

Antifoaming agents like polyethylene glycol are not considered toxic and low molecular masses are generally biodegradable to a high extend, e.g. about 80 % of the fraction with a molecular mass of less than 4000 g/mole (Falbe and Regitz, 1995). As polyethylene glycol can be highly polymerized up to 5,000,000 g/mole, some compounds may be rather persistent in the marine environment. With a discharge concentration of 0.05 mg/l (Table 3.6 on page 39), a large MSF plant like Doha West will add about 180 kg organic antifoamings to the coastal water each day (Table 4.1). With regard to their non-toxicity and relatively low discharge levels, antifoaming agents are of little concern for the marine environment and will increase natural background concentrations of dissolved organic carbon only slightly in the mixing zone. Nevertheless, polypropylene glycol and polyethylene glycol were replaced in some industrial applications for their poor biodegradability by dialkyl ethers (R_1-O-R_2, Falbe and Regitz, 1995), which might be used in desalination plants as well to improve biodegradability of antifoaming agents.

4.7 Corrosion

4.7.1 Copper

Heavy metals are claimed to be a major pollutant of desalination discharge and are probably increased during acid cleaning of MSF distillers (Mannaa, 1994; Shams El Din et al., 1994; Abdel-Jawad and Al-Tabtabaei, 1999; El-Dessouky, 2000). However, no information on heavy metal concentrations during cleaning is available and levels in brine blow-down from normal operation are rarely published, although the copper content is usually monitored to detect pitting and crevice corrosion of heat exchanger alloys.

Oldfield and Todd (1996) reviewed copper contents in brine blow-down and pointed out, that levels will be less than 100 ppb in modern plants with good process control. For example, an acid treated plant discharged 200 ppb in 1979, which could be considerably reduced by treatment with polymer antiscalants. The lowest reported level in brine blow-down was an average of 15 ppb in the MSF plant in Al Jubail, which is according to Mannaa (1994) "well below the water quality criterion of 23 μg/l set by the EPA" and consequently considered as "insignificant to induce deleterious effects on the local ecosystem". The author refers to the US Environmental Protection Agency's ambient water quality criteria (WQC) from 1980, however, these criteria specified a permissible concentrations of total recoverable copper in saltwater of "4.0 μg/l as a 24 hr average", which "should not exceed 23 μg/l at any time" (HSDB 2001p). EPA standards have become more rigorous since then and take the high toxicity of dissolved copper into account by setting 4.8 μg dissolved copper per liter as a maximum concentration for brief exposure (CMC) and 3.1 μg/l for long-term exposure (EPA, 1998). For comparison, the latter value is more conservative but still in agreement with a predicted no effect concentration (PNEC) for European saltwater of 5.6 μg/l (Paquin et al., 1999). Therefore, even a minimum discharge of 15 ppb in the brine blow-down exceeds the chronic criterion by a factor of 5 and does not comply with the revised acute criterion of 4.8 μg/l any longer. This minimum as well as higher brine concentrations up to 100 ppb are consequently only acceptable, if further dilution occurs prior to discharge.

The MSF brine is usually blended with about twice the amount of cooling water and this factor of dilution is commonly assumed for the effluent concentration of copper, giving a theoretical range of 5-33 ppb in the combined effluent (based on Oldfield and Todd, 1996; Mannaa, 1994). However, the current practice of dilution must be questioned, as the rejected cooling water could contain traces of copper from copper-based tubing material, too. Seawater flow rates are high and the water is not deaerated in the heat rejection section, which is only done for the feed water to the heat recovery section (Fig. 2.2). As failure of copper-based alloys is often due to residual oxygen, which causes local erosion-corrosion or pitting (Oldfield and Todd, 1996), the cooling water is presumably not free from copper even though more resistant 70-30 cupro-nickel is used for construction in this section of the plant (cf. section 2.2.1 on page 18). For comparison, Paquin et al. (1999) reviewed copper in cooling water of power plants and observed levels between 1-4 μg/l. Irrespective of the unknown amount of copper in the cooling water, it must be concluded that the current practice of dilution still yields levels which are above the short-term as well as long-term permissible concentrations for seawater (5-33 μg/l compared to 4.8 and 3.1 μg/l). Mixing in the receiving water body will therefore adjust concentrations to ambient levels.

Copper can also be enriched in the environment naturally, e.g. through copper-rich terrestrial material, so that the mentioned WQC are not always applicable. For example, concentrations range from an oceanic background of about 0.1 μg/l to 100 μg/l in estuaries (Kennish, 1997), and the latter value exceeds the mentioned WQC by a factor of 20. Natural background levels have to be considered for the assessment of copper in desalination plant discharge. However, it is difficult to distinguish between natural and anthropogenic-influenced copper levels in the Arabian Gulf, as they result from sewage, industrial outfalls, off-shore oil exploitation and as a consequence of the Gulf War oil spills (UNEP, 1999; Sadiq, 1996). Higher levels of copper have generally been found in coastal areas close to ports and heavy industry but decreased further off-shore. A gradual decline from 3.4 μg/l to 0.6 μg/l was for example observed on a transect from Doha, Qatar, to open Gulf water (Abdel-Moati and Kureishy, 1997)°. The estimated discharge concentrations of desalination plants are therefore well within the range that could contribute to an increase of coastal copper concentrations, and water quality criteria may be exceeded in the mixing zone of desalination effluents.

Toxicity of released copper depends on site-specific conditions of pH and dissolved organic carbon which influence metal speciation and bioavailability. Copper ions are more soluble in acidic waters while alkaline conditions favor precipitation, e.g. as cupric carbonate or hydroxide (HSDB 2001q), with seawater being slightly alkaline. Dissolved copper is also prone to complex formation with humic substances and complexes are more easily adsorbed by clay or other suspended material. Thereby, dissolved amounts of copper can be decreased considerably while the element is enriched in the solid phase and finally in sediments (HSDB 2001r). As organic and inor-

° Original values were 53.2-9.8 nM/kg.

ganic metal complexes are not bioavailable, toxicity is reduced and this may be taken into account by implementing site-specific water quality criteria (Paquin et al., 1999). Although these criteria will be higher and more easily met by desalination and power plant effluents, it is evident that the problem is transferred from the water column to the sediment. Consequently, the discussion of water concentrations does not cope with the problem entirely and the current practice of waste stream blending must be objected with the argument, that it is the load rather than the concentration which poses a threat to the marine environment (Hoepner, 1999). Most of this extra material will enter the sediments, and deposition of unacceptable loads of effluent constituents should be prevented, e.g. as prohibited according to EPA regulations.

However, the mere presence of copper does not necessarily mean that it will adversely effect the marine environment. As copper is an essential micro-nutrient for most organisms, it is accumulated and might only become toxic, if excess amounts are biologically available, depending on physiological condition and general species sensitivity. Organisms may also adapt to sublethal concentrations by binding copper, which reduces the freely available concentration in the body (Fent, 1998). Due to metal speciation on the one hand and a gradual shift between necessary uptake and toxicity on the other, the impact of copper on the marine environment can hardly be determined. This is in line with the frequent observation, that field copper levels associated with a certain effect are often higher than levels determined in the laboratory (Paquin et al., 1999), i.e. the mentioned environmental reactions diminish copper toxicity. Some results of toxicity tests are provided in Fig. 4.6, which show that discharge concentrations of desalination plants could have acute effects on more sensitive species, while others will be more resistant, e.g. by adapting to sublethal concentrations. However, chronic effects are nevertheless possible. Although no information is available for acid cleaning, discharged concentrations are probably higher and more bioavailable due to lower pH, which increases the risk for acutely toxic situations.

4.7.2 Nickel

Copper-nickel heat exchangers contain 10 % or 30 % nickel and traces of this element will pass into the brine, too. As no data is available on dis-

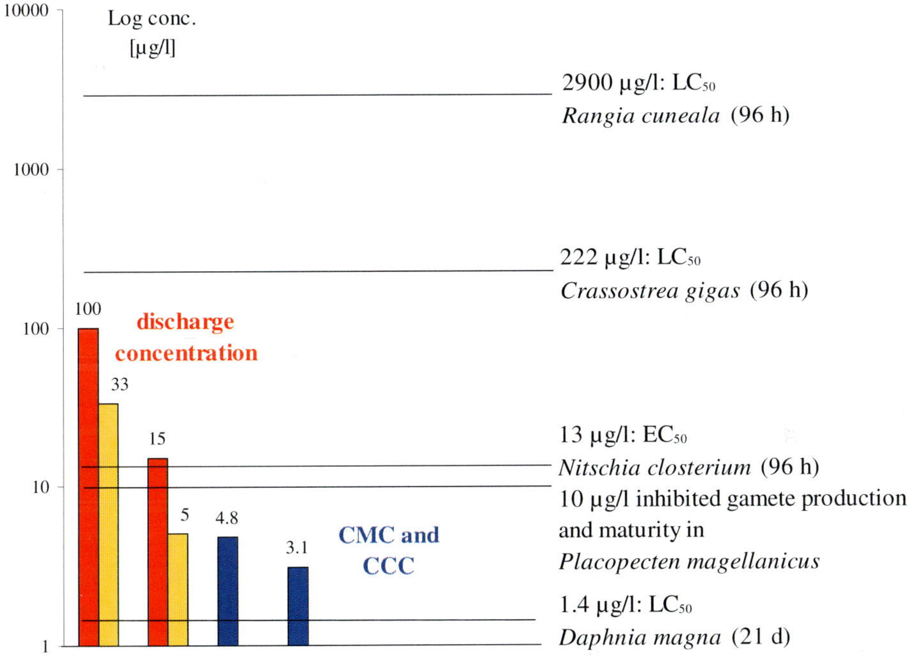

FIGURE 4.6: Comparison of copper concentrations in the brine (15-100 μg/l) and diluted desalination discharge (5-33 μg/l) with water quality criteria (CMC and CCC) and toxicological data. All data refer to saltwater species and are derived from tests with copper sulfate and 96 h exposure, if not specified differently. For a better comparison with copper discharge, the respective copper concentrations were calculated based on the assumption, that sulfate is of less concern for saltwater organisms. The original concentrations are as follows: Effect concentration EC_{50} on growth rate of the algae *Nitschia closterium* was 33 μg/l (HSDB 2001s). Molluscs withstood higher levels like the oyster *Crassostrea gigas*, LC_{50} of 560 μg/l (HSDB 2001t), or clams of *Rangia cuneala*, 7400 μg/l (HSDB 2001u). Long-term tests conducted on *Daphnia magna* (freshwater) lasted 21 days and produced an LC_{50} of 1.4 μg copper/l (HSDB 2001v) and chronic effects were observed in the range of 10-20 μg copper/l/8 weeks, which inhibited gamete production and maturity of sea scallops *Placopecten magellanicus* (HSDB 2001w).

charge levels, Oldfield and Todd (1996) estimated the nickel content in the brine to be in proportion to its alloy concentration and to the rate of copper dissolution. This approach does not take denickelification into account (cf. section 3.5 on page 38), which is difficult to quantify. Assuming again a minimum of 15 μg/l copper in the brine, nickel concentrations might be the same but nickel is generally assumed to be less toxic than copper. For example, the 21 day toxicity test of *Daphnia magna* produced a considerably higher LC_{50} of 130 μg nickel ion/l (HSDB 2001x). The lower toxicity of nickel components is reflected in higher water quality criteria recommended by the US EPA, which are 74 μg/l as a maximum concentration and 8.2 μg/l for long-term exposure (EPA, 1998). Typical seawater concentrations of nickel, mainly Ni^{2+}, chloro- and carbonato-complexes, are between 0.1-0.5 μg/l (HSDB 2001y), which is well below the regulatory long-term level. However, background data could be higher in the Gulf. Nickel is an oil-related metal (Fowler et al., 1993), so that oil contamination is a major source of nickel input. With regard to the available data, it can be concluded that nickel concentrations in the brine comply with the short-term criterion for exposure and could meet the chronic criterion by dilution with cooling water, depending on its nickel content. Although dilution in the coastal water body will adjust levels to ambient, it is again the environmental fate that determines total amounts in the vicinity of plant outfalls.

Similarly to copper, the mobility of nickel is increased at low pH values (<6.5, HSDB 2001y) and therefore higher in acid cleaning wastes. Fish and marine plants can accumulate soluble nickel species but the potential for bioconcentration is low to moderate (HSDB 2001x). Furthermore, complex formation with organic substances, scavenging by suspended solids and co-precipitation with iron and manganese hydroxides (HSDB 2001z & r) decrease the availability of dissolved nickel and enhance deposition in sediments. Although nickel is one of the most mobile of the heavy metals in the aquatic environment, considerable amounts comparable to copper might be put down in the sediments near desalination plants. As long as environmental data is sparse, environmental effects of different nickel compounds cannot be excluded.

4.7.3 Iron, chromium and molybdenum

Discharge information from nickel-copper alloys is already rare, but it is virtually non-existent for heavy metals from stainless steel construction material in desalination plants (i.e. Fe, Cr, Ni and Mo). General corrosion is usually very low but pitting and crevice corrosion is possible if process control mechanisms fail. Oldfield and Todd (1996) calculated quantities which continuously pass into the brine as a result of passive film formation. According to these, iron is discharged at levels of 0.13-13 μg/l depending on passive current, which is less than the brine contamination of 25 μg/l mentioned by Mannaa (1994). No EPA saltwater quality criteria exist for iron and it is considered a non-priority pollutant. Calculated chromium concentrations in the discharge are 0.035-3.5 μg/l, which are well below the maximum criterion of 1,100 μg/l and the long-term criterion of 50 μg/l for chromium VI (EPA, 1998), which is the species of toxicological significance (Falbe and Regitz, 1995). Molybdenum concentrations were calculated to be 0.004-0.4 μg/l, an extremely low value as molybdates have a low toxicity and no EPA WQC are specified. For nickel, concentrations are assumed to be lowest with 0.002-2 μg/l and these are also below the above mentioned WQC.

4.7.4 Heavy metal assessment

To conclude, copper-nickel alloys are considered problematic predominantly due to copper discharge. It should be noted that the previous discussion of copper and nickel is mostly based on one single value (15 μg/l) which is the lowest obtainable estimate for copper discharge. Effluent concentrations could be higher, but information on brine and cooling water contamination during normal operation as well as during acid cleaning is incomplete for both metals. Toxicity of heavy metals depends on the metal compound present under environmental conditions, and site-specific criteria could be developed which will probably be less stringent than general criteria. The same processes that reduce toxicity also favor the heavy metal transfer from water to sediment and special attention should be given to total sediment loads. This point of discussion is therefore resumed in section 5.2 as one aspect of pollution in the Arabian Gulf.

Copper-nickel alloys can partly be replaced by titanium tube material to diminish copper input, but is restricted to stages of low brine temperature, i.e. less than 80 °C (El-Dessouky, 2000). For example, both materials are implemented in the Taweelah B plant in the UAE with 345,000 m^3 capacity, which started operating in 1995 (Fisia Italimpianti, 2001; Wangnick, 1999). In contrast to copper-nickel heat exchanger tubes, less metal seems to pass into the brine from stainless steel construction material. Actual measurements in the brine and studies on accumulation in sediments or organisms are still missing to support this thesis, but heavy metal contamination through RO plants is probably of less concern compared to MSF plants.

4.7.5 Benzotriazole derivates

Corrosion inhibitors like benzotriazole and its derivates are necessary to protect copper based condenser tubes during acid cleaning in MSF plants. Although they are thought to form a stable film by interaction of their unshared electrons with metallic surfaces (cf. section 3.5), it must be assumed that most of this treatment chemical is finally discharged to the sea. Benzotriazole is rather persistent in surface water as biotic and abiotic degradation is slow and adsorption to particles or volatilization to the atmosphere are minor fate processes. However, the tendency for adsorption is strongly increased in acidic environment when benzotriazole is dominant in the protonated form (HSDB, 2001). Therefore, the inhibitor might adsorb to suspended material and accumulate in sediments due to the low pH of the cleaning discharge, where it is slowly degraded.

Discharge amounts are difficult to estimate as brine concentrations of benzotriazole are not obtainable. If 25 mg/l dosing of inhibitor (Table 3.7 on page 41) is representative, a total amount of 125 kg per distiller or 2000 kg is required for cleaning a large plant like Doha West[p] (Table 4.1). This amount does not enter the coastal water at once as most molecules will adhere to the plant's interior surfaces. It is also unlikely that all distillers are cleaned simultaneously, otherwise the plant had to go off-line and stop production. Furthermore, cleaning waste might be blended with brine discharge so that environmental concentrations are not easily predictable with

[p] Assuming 5000 m^3 cleaning volume per distiller or a total of 80,000 m^3 for 16 distillers.

the given information. Toxicological information rarely exists, but acutely toxic effects are not to be expected for benzotriazole, as LC_{50} of trout and *Daphnia magna* are relatively high with 39 mg/l/96 h and 141,6 mg/l/48 h respectively (pers. communication Knoll, 2001) and exceed the dosing concentration sufficiently. The potential for bioconcentration in aquatic organisms is low (HSDB, 2001), yet, the substance is classified as harmful to aquatic organisms according to the environmental data provided by ICSC (2000), which may be due to its longevity and possible chronic effects.

If condenser tubes were replaced by more resistant construction materials (e.g. plastic, El-Dessouky, 2000), dosage of corrosion inhibitors could probably be reduced or might become obsolete. Another favorable material is titanium but inhibition appears to be necessary during cleaning to avoid hydrogen adsorption (Andijani et al., 2000). As long as copper-nickels are common materials for heat exchanger surfaces, corrosion inhibitors will not only protect the plant from corrosion but will also minimize the risk of heavy metal pollution, although the impact by the inhibitor can hardly be assessed.

4.8 Membrane cleaning chemicals

4.8.1 Formaldehyde

If formaldehyde is released into water, it will be further diluted and is usually biodegraded within a few days (HSDB 2001aa). Therefore, an impact on the environment may be noticeable immediately after membrane cleaning when initial concentrations are still high. Other processes like photolytic degradation or volatilization from water contribute only slightly to the overall decrease of environmental levels (HSDB 2001bb & cc). These levels are difficult to estimate, as little is known on the current practice of cleaning waste disposal but recommended dosage of cleaning chemicals is available from membrane manufacturers.

For example, DuPont (1994) specifies a concentration of 0.5-1 % formaldehyde solution for disinfection, i.e. 5-10 ml/l, which amounts to 2920-5840 l formaldehyde for an RO plant the size of Al Jubail (cf. Table 4.2 on page 81). If the waste was disposed of by discharge to the sea, dilution in 29,000-58,000 m^3 seawater will yield a concentration of 0.1 ml/l and this level must still be considered acutely toxic for some marine organisms. The

TABLE 4.1: Doha West MSF plant: properties and chemicals

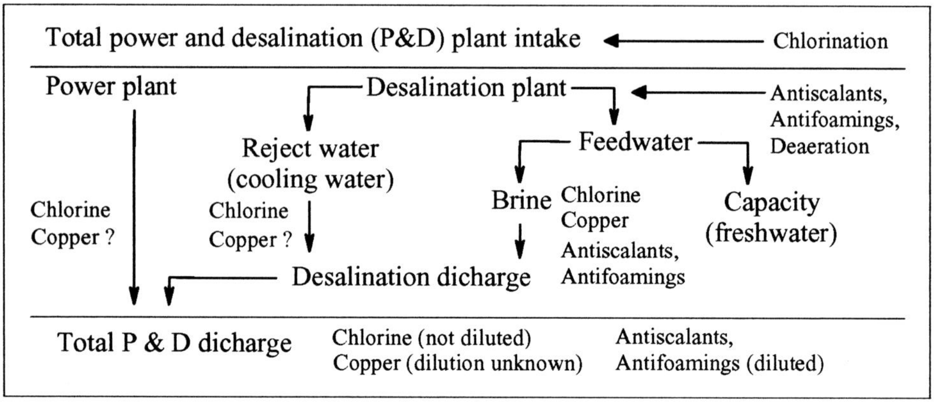

Plant properties	**Volumes [m^3]**		**Salinity**	**Temperature [°C]**
Total P&D intake	10,483,200		45	33
1. P plant intake	6,336,000			
2. D plant intake	4,147,200	100 %		
2.1. Feedwater	1,244,160	30 %	ambient	
2.1.1. Capacity	436,465	10.5 %	–	
2.1.2. Brine	807,695	19.5 %	69	
2.2. Reject water	2,903,040	70.0 %	ambient	
D discharge	3,710,735	89.5 %	50	
Total P&D discharge	10,046,735		47	38

Pretreatment chemicals and dosage [ppb]	**Effluent concentrations [ppb] −desalination**	**−total P&D**	**Load/day [kg]**	**Water quality criteria [ppb]**	
Chlorine (2500)	250	250	2,512	13	CMC
Chlorine (5000)	500	500	5,023	7.5	CCC
Polymer (2000)	671	248	2,488	580,000	LC_{50}
Antifoam	50	18	186	–	
Copper (brine: 15)	?	?	12	4.8	CMC
Copper (brine: 100)	?	?	81	3.1	CCC

Acid cleaning	**Cleaning solution [m^3]**	**Seawater [m^3] required for neutralization**	**Load [kg]**	**Water quality criteria [ppb]**	
Benzotriazole					
1 unit (25,000 ppb)	5,000	20,000	125	39,000	LC_{50}
16 units (25,000 ppb)	80,000	320,000	2,000		

Plant properties based on Abdel-Jawad and Al-Tabtabaei (1999): The desalination plant receives on average 1,600 m^3 seawater per hour for an installed capacity of 1 MIGD and about 100 m^3/h per MW. Doha produces 96 MIGD (436,465 m^3/day based on imperial gallons) and 2400 MW. In dual purpose plants, the brine is usually blended with cooling water, so that salinity as well as some residual chemical concentrations are decreased in the combined effluent.

TABLE 4.2: Al Jubail RO plant: properties and chemicals

		Salinity		
Plant properties	**Volumes [m^3]**	**average**	**min**	**max**
Intake	288,000	43.8	42.0	46.5
Capacity	91,000	–	–	–
Conversion %	0.32			
Brine	197,000	64.0	61.4	68.0
Mixing volume 1:10	2,166,998	45.6	43.8	48.5
Pretreatment chemicals and dosage [ppm]	**Load/day [kg] or [l]**	**Concentration after dilution [ppm] −1:10**	**−in 125,000 m^3**	
Chlorine	neutralized	–		
FeO(OH)	50	0.02	0.4	
Polyelectrolyte (0.3)	86	0.04	0.7	
Suspended Matter (1)	288	0.13	2.3	
SHMP (2)	576	0.04		
Polymer (2)	576	0.04		
Cleaning chemicals				LC_{50} **[mg/l] or [μl/l]**
HCl (0.5V%)	2920	neutralized	23.4	
$C_6H_8O_7$ (2W%)	11680	neutralized	93.4	
H_3PO_4 (0.5W%)	2920	neutralized	23.4	
NaOH (0.1V%)	584	neutralized	4.7	
SHMP (1W%)	5840	2.7	46.7	
NaDSS (0.025W%)	146	0.1	1.2	1
NaDBS (0.25W%)	1460	0.7	11.7	?
$NaBO_3$ (0.3W%)	1752	0.8	14.0	11
STP(2W%)	11680	5.4	93.4	?
Na – EDTA(1W%)	5840	2.7	46.7	10-100
TSP (1W%)	5840	2.7	46.7	?
Formaldehyde (1V%)	5840	2.7	46.7	100
Glutaraldehyde (1W%)	5840	2.7	46.7	?
Isothiazole (1.5%, 20 ppm)	0.18	0.0001	0.001	0.2

Plant properties based on Baig and Kutbi (1998). Load calculation for solid chemicals: load [kg] = W% · 10 [g/l] · V [l]; for fluids: load [l] = V% · 0.01 · V [l] with V = 584,000 liters (cf. section 2.1.4). W% and V% assume each chemical is added as 100 % active ingredient (DuPont, 1994). Dilution 1:10 is assumed for those chemicals, that might be blended into the desalination discharge. The dilution factor is derived from salinity levels, i.e. the factor that is necessary to decrease salinity close to ambient. Dilution in a theoretical impact area of 125,000 m^3 is assumed for cleaning and filter waste material, if it was disposed of all at once. It should be noted, that waste dilution is prohibited in some countries and no means of waste disposal. $C_6H_8O_7$ citric acid, SHMP sodium hexametaphosphate, NaDSS sodium dodecylsulfate, NaDBS sodium dodecylbenzene sulfonate, $NaBO_3$ sodium perborate, NaOCl sodium hypochlorite, STP sodium triphosphate ($Na_5P_3O_{10}$), NaEDTA sodium ethylenediamine tetraacetic acid, TSP trisodium phosphate (Na_3PO_4), SBS sodium bisulfite.

assumption is based on the LC_{50} of bluegill, which was equal to 0.1 ml/l for 96 h exposure (HSDB 20011), and the fact that biocides like formaldehyde are prohibited for discharge in the USA (EPA, 1999). For illustration, the calculated volumes equal a theoretical mixing zone of 125,000 m^2 with a maximum depth of about 0.5 m.

For mitigation, formaldehyde can be deactivated with H_2O_2 and $Ca(OH)_2$ or NaOH prior to discharge, which is usually the case for wastewater treatment (Falbe and Regitz, 1995). Mickley (1995) mentions continuous blending of cleaning disinfectants with the brine to minimize impacts but dilution is generally no means of waste disposal. Glutaraldehyde may be used alternatively as a biocide and should be applied in the same concentration as formaldehyde. This replacement is also alarming as glutaraldehyde is classified as very toxic to aquatic organisms according to international chemical safety cards (ICSC, 2000).

4.8.2 Isothiazole

Isothiazole derivates are highly effective biocides and therefore also very toxic to aquatic organisms if released to the marine environment. They are recommended by some membrane suppliers for disinfection and storage (Trisep, 1999; Hydranautics, 2000) or might be added to the feedwater continuously (BetzDearborn, 2000). Cleaning waste disposal must usually comply with special regulations, but residual isothiazole in the brine from normal operation will probably enter the coastal water untreated. However, no discharge has been reported from desalination plants so far.

To disinfect a membrane, a commercial solution with 1.5 % active ingredient can be added at 20 ppm (Hydranautics, 2000). A commercial product with comparable concentration of 0.6-3 % 5-chloro-2-methyl-isothiazole-3-one and 0.06-0.6 % 2-methyl-isothiazole-3-one, produced the following LC_{50} values: *Daphnia magna* 2.9 mg/l/48 h, bluegill 12.1 mg/l/96 h, sheepshead minnow 20 mg/l/96 h (BetzDearborn, 2000). As these results from acute tests are all below the recommended sanitizing dosage of 20 mg/l, accidental or intentional discharge might have an impact on marine organisms in the vicinity of the outlet before the chemical is diluted sufficiently. However, the affected area will be comparatively small as low volumes of cleaning solution are required in RO plants. For bluegill, dilution with the same amount of seawater will suffice to decrease the environmental concentration below the no observed effect limit, which is 6.5 mg/l.

For comparison with formaldehyde, the same model calculation may be helpful based on the example of Al Jubail. If 20 mg/l commercial product (1.5 %) was added to a disinfection volume of 584 m^3, the active ingredient would amount to about 0.2 kg (Table 4.2). Although both disinfectants must be considered as similarly toxic to bluegill[q], a lower dosage of isothiazole is recommended. It has to be expected that the overall impact will be less if this biocide is discharged to the sea. Subsequent dilution will account for a decrease in environmental concentrations as the product was classified as not spontaneously biodegradable. With regard to the given information, storing and continuous blending of wastes with the brine as suggested by Mickley (1995) would further minimize the risk of toxic effects. However, the chemical waste can also be treated with sodium bisulfite for neutralization prior to discharge (Trisep, 1999) if disposal of biocides is generally not permitted in coastal waters.

4.8.3 Sodium dodecylbenzene sulfonate

Sodium dodecylbenzene sulfonate (Na-DBS) may become a waste product after cleaning of RO membranes and is probably released into the coastal environment. As volatilization and adsorption to sediment are minor fate processes, a decrease of environmental concentration is most likely due to dilution and biodegradation. Several biodegradation studies from HSDB (2001) indicate that the hydrophilic sulfonate group is removed (yielding sulfate) before the linear alkyl-chain is degraded by oxidation. The lipophilic intermediate products still contained the aromatic ring structure[r] but were almost completely decomposed after a certain time (HSDB 2001ee). According to Falbe and Regitz (1995), Na-DBS has become an important anionic detergent for its high biodegradability, and surfactants with linear aliphatic residues are generally more degradable than branched ones. Although environmental data is rare, Zahn-Wellens tests indicate that concentrations of sodium dodecylbenzene sulfonate were reduced by

[q] The LC_{50} of 12.1 mg/l commercial product corresponds to about 0.2 mg/l active ingredient, if the product contains 1.5 % isothiazole. This has to be compared to an LC_{50} of 0.1 ml/l formaldehyde (100 % active ingredient).

[r] e.g. 1-tetralone, 1-indanone, 4-methyl-1-tetralone and naphthalene (HSDB 2001dd)

87 % in 17 days (HSDB 2001ff) and the rate of degradation was found to be highest in 25-30 °C warm water (HSDB 2001gg), i.e. in subtropical regions where desalination plants are predominant.

Dosage of this detergent is recommended in a concentration of 0.25 % for cleaning (DuPont, 1994), so that a plant like Al Jubail will theoretically discharge 1,460 kg (Table 4.2). If this amount was blended with the daily discharge stream and diluted by a factor of ten, as necessary to reduce brine salinity to about ambient, the discharge concentration would be about 0.7 mg/l. In contrast, dumping of the whole amount into a theoretical mixing zone (125,000 m^2 and 1 m depth) would yield a concentration of 11.7 mg/l. No toxicological information was obtainable for comparison but toxicity of detergents is usually high and depends on exactly this property, which disturbs the intracellular membrane system of organisms (Hoepner, 1999). As primary degradation removes the hydrophilic group from the lipophilic chain, toxicity is expected to be relatively low once the components have been separated. The intermediate products may become enriched in fatty tissue of organisms due to their lipophilic properties, but nothing has been reported so far and bioconcentration is not to be expected (HSDB, 2001). This might be attributed to the substance's short residence time in the marine environment as the alkyl-chain residue is biologically degradable. Nevertheless, surplus organic material is added to the coastal water, which must be compared with background levels of 3.5 mg/l total organic carbon near Al Jubail. Although cleaning is only necessary in certain time intervals, this extra load must be degraded in the coastal water body with a source of residual chlorine nearby. In Al Jubail, an installed MSF capacity of 1,15 million m^3 per day (Wangnick, 1999) will result in chlorinated by-products, which are formed depending on the availability of organic seawater constituents.

4.8.4 Sodium dodecylsulfate

Sodium dodecylsulfate (Na-DSS) is recommended by FilmTec (2000) as anionic detergent for biofouling control and its environmental fate is similar to Na-DBS characterized by biodegradation. Rates varied considerably in surface water samples and were between 2-6 days for a reduction by 80 % of the initial concentration (25 mg/l, test conditions not specified, HSDB 2001hh). Recommended dosage was by a factor of ten lower than for Na-

DBS, and a concentration of 0.025 % will amount to a total discharge of 146 kg at Al Jubail, which corresponds to a concentration of 1.2 mg/l in a theoretical impact area of 125,000 m^3 or 0.07 mg/l if the waste was blended with the daily discharge from the desalination plant (Table 4.2). In contrast to Na-DBS, toxicological information was obtainable and LC_{50} ranged between 1-10 mg/l for fish, *Daphnia magna* and algae (Redox-Chemicals, 2001). These relatively low values are in line with the general classification as a toxic substance for aquatic organisms by international chemical safety cards (ICSC, 2000). However, hydrolysis of the sulfate bonding will again remove the hydrophilic group and probably reduce overall toxicity of the intermediates until degradation is complete. Although the lipophilic products are likely to be accumulated, rapid degradation reduces the exposure time of organisms to these substances. By comparison of toxicity data and estimated environmental concentrations, it becomes evident that the practice of dilution decides about the marine impact, as always. As an alternative, microbial degradation could destroy the surface active property and degrade the residual alkyl-chain in waste treatment tanks prior to discharge. This option is preferable from an environmental point of view, as dilution is generally not an adequate means of waste disposal.

4.8.5 Ethylenediamine tetraacetic acid

EDTA and corresponding sodium salts are used in RO cleaning solutions for removal of inorganic colloids and biofouling, as they provide effective chelating and dispersing properties. EDTA is recommended by most membrane manufacturers in varying concentrations and in different combinations, as discussed in section 3.6 and shown in Table 3.8. For the Dupont B-10 permeators in Al Jubail, a dosage of 10 mg/l (equal to 1 %) would result in a total discharge of 5,840 kg. Depending on the practice of disposal, 2.7 mg EDTA/l has to be expected if the waste load is blended with the discharge or a concentration of 47 mg/l is estimated for dilution in a seawater volume of 125,000 m^3 (Table 4.2).

Residual EDTA is problematic in the coastal water body as biodegradation is slow, e.g. a reduction by less than 5 % was observed after 21 days when 20 mg/l EDTA were inoculated with activated sludge under aerobic conditions (HSDB 2001ii). In addition, observations of microbial activity in a lagoon indicate that major biodegradation products are ethylenediamine triacetic acid, ethylenediamine diacetic acid or nitrilotriacetic acid (HSDB 2001jj), which are also chelating agents. In the environment, EDTA complexes can contain trace metals like Cu, Ni, Zn, Cd, Mn, Fe or alkaline earth metals (Ca, Mg), but calcium chelates are predominant under alkaline conditions (HSDB 2001kk), i.e. in seawater. While complex formation increases overall metal solubility and mobility on the one hand (HSDB 2001ll), metal complexes are comparatively non-reactive on the other hand (Falbe and Regitz, 1995) and thereby metal bioavailability is reduced. The chelating properties of EDTA might interfere significantly with natural cycles of these elements in coastal waters or could even re-mobilize metals from sediments. As degradation, volatilization from water or deposition in sediments are minor fate processes of EDTA complexes (HSDB, 2001), the major decrease in concentration will be due to dilution. Bioaccumulation is also negligible[s], so that toxicity will depend on dispersion of the discharged material in seawater. An LC_{50} for bluegill (*Lepomis macrochirus*) of 159 mg/l was specified for 96 h exposure and the no observed effect level for this time span was 100 mg/l (HSDB 2001l). This is higher than the given information by Trisep (1999) for one of the company's commercial cleaning products, although it is less concentrated. The product contained up to 60 % Na_2EDTA and produced acute effects in the range of 10-100 mg/l, which are as follows: *Daphnia magna* $EC_{50} > 100$ mg/l/24 h, Golden Orfe $LC_{50} > 50$ mg/l/96 h, algae EC_{50} 10-100 mg/l/72 h. With regard to the given information, it can be concluded that the impact of EDTA discharge does not depend on direct toxicity. Most obtainable LC_{50} values are above the calculated maximum estimate of 46 mg/l, which can still be considerably reduced by blending. For EDTA, total loads are more of concern due to the chemical's persistence in the environment, in which its dispersing and chelating properties will not cease to exist and could influence reactions in the water column.

4.8.6 Sodium perborate

$NaBO_3$ is recommended by DuPont (1994) as a cleaning agent for oxidizing organic deposits on the membrane, and a dosage of 0.3 % is considered effective for this purpose. This amounts to 1,752 kg or a concentration of

[s] A bioconcentration factor for EDTA in bluegill was measured to be < 2 (HSDB 2001mm).

14 mg/l if this load is projected to a volume of 125,000 m^3 (Table 4.2). However, $NaBO_3$ decomposes to sodium borate and H_2O_2 in water, and H_2O_2 is the active ingredient which is used for its oxidizing properties by setting in turn oxygen free. The oxidizing potential is reduced by reaction with organic films on the membrane, and the amount of residual H_2O_2 which could damage marine life in the environment can hardly be determined. Sodium perborate is generally considered as non-hazardous, which is due to the relatively weak biocidal effect of H_2O_2 compared to chlorine and its self-decomposition to water and oxygen. For comparison, acute tests with $NaBO_3 \cdot H_2O$ produced relatively high LC_{50} of 11 mg/l/48 h (*Daphnia magna*), a fish toxicity at 51 mg/l/96 h (*Brachydanio rerio*) and an algae toxicity (EC_{50}) at 12 mg/l/96 h (*Scenedesmus suspicatus*). The no observed effect concentrations were 8 and 25 mg/l for zooplankton and fish, respectively (Redox-Chemicals, 2001). Under the assumption that residual oxidant concentration are considerably lower than 14 mg/l, acute effects are not very probable even for more sensitive species. This is in line with other studies that generally produced higher values, e.g. 160-320 mg/l/24 h for fingerling trout, on which a commercial membrane cleaning product was tested that contains 90 % sodium perborate (Trisep, 1999). Environmental effects of sodium perborate were reported as concentrations exceeded 250 mg/l in sewage, which harmed brackish water organisms (HSDB 2001nn). To allow for sufficient self-decomposition, the waste material could be temporarily stored prior to discharge.

4.9 Impact assessment

In the previous sections, single constituents of the desalination plant waste were assessed with regard to their likely environmental impacts. By comparison of these results, it can be concluded that normal operation of RO plants will be of less concern for the marine environment than the continuous discharge from MSF plants. In contrast, cleaning solutions that are periodically used to restore the performance of RO membranes are generally more hazardous than the waste produced during the MSF cleaning procedure. The arguments that support this thesis are summarized in the following sections. Furthermore, the purpose of a plant is manifested in plant size and preferred location, which are different for the two de-

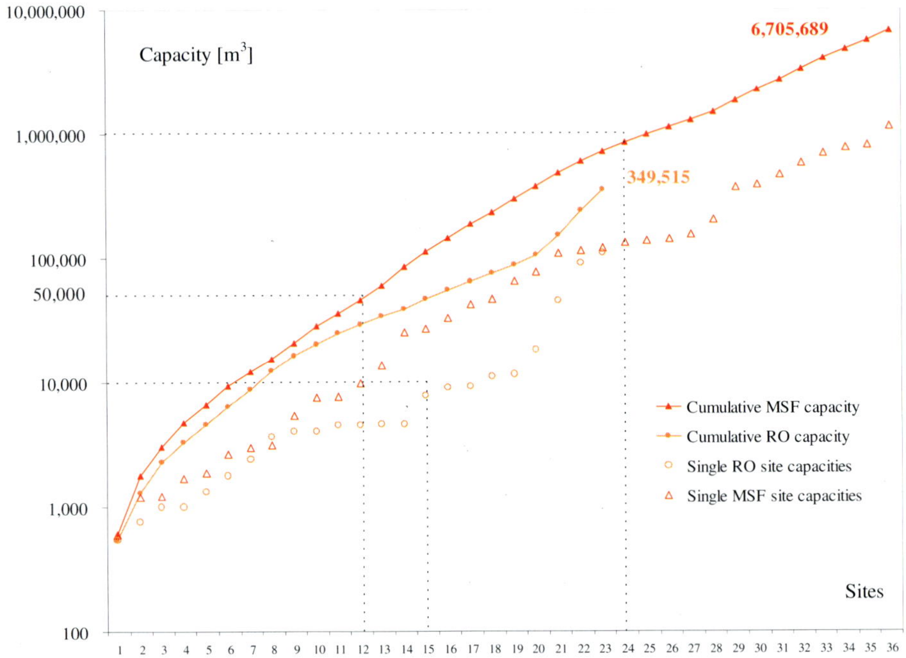

FIGURE 4.7: Cumulative capacities of MSF and RO plants in the Arabian Gulf: Two thirds of all RO plant sites (15 out of 23) have an installed production of less than 10,000 m^3 and a cumulative capacity of 50,000 m^3. The same trend is observable in MSF capacities, but a cumulative capacity of 50,000 m^3 is achieved by one third of all sites (12 out of 36), whereas another third accounts for 80 % of the total MSF production. Data source: Wangnick (1999).

salination processes. These parameters are per se important factors that determine the impact on the marine environment and will be considered in addition to the chemical aspects of the discharge.

Plant size and location

RO plants are generally smaller than MSF plants. Although single RO plants can have an installed capacity of about 100,000 m^3 per day, most plants produce considerably less potable water: For example two thirds of all plant sites in the Arabian Gulf have an installed production of less than 10,000 m^3 per day, and their cumulative capacity is only 50,000 m^3 (Fig. 4.7). In contrast, the trend towards large plants is more pronounced for the MSF process, especially in the oil-rich countries bordering the Gulf: Again, a cumulative capacity of 50,000 m^3 is achieved by plants smaller than 10,000 m^3, but these account for only one third of all sites. Another third, however, surpasses the upper RO limit of 100,000 m^3 per day, and accounts for roughly 5 million m^3 or more than 80 % of the total MSF production.

Lower energy consumption of RO plants is an important factor, which makes the process more favorable for decentralized development in coastal areas. Small towns, tourist areas or single hotels have a low to moderate water demand, and might account for an increase in RO plant numbers in the future. This trend has probably no marked influence on total capacities, which are dominated by large thermal plants that often arise close to urban areas, industrial complexes or power stations. It is difficult to assess whether the concentration of high capacities in some places is more acceptable from an environmental point of view or the spreading of smaller facilities throughout large coastal areas. With regard to each single plant, impacts on the environment will depend on plant size but also on the sensitivity of the ecosystem which receives the brine discharge (Höpner and Windelberg, 1996). A major disadvantage of decentralization is therefore that plants may be situated in relatively pristine environments. These are at risk from the effluent, although returned volumes are comparatively low. Blending with other waste streams is therefore recommended to reduce brine salinity or alternatively, evaporation ponds could solve the problem for small plants (Ahmed et al., 2000). In contrast, MSF plants might represent only one source of pollution in a coastal water body that is multiply used, e.g. by shipping and industries. High discharge volumes will further increase the stress on the marine environment. Although the trend is towards different use of the two processes, it can be concluded that both may have adverse effects on the receiving water body. Waste stream disposal must comply with environmental standards independent of plant size and location. However, standards can be more or less stringent, thereby accepting that human activities deteriorate water quality to a certain degree in some places (GESAMP, 1986). On the other hand, this classification scheme has to give high priority to intact nature. While multiple sources of pollution might be more of concern for MSF plants, the discharge of a single plant could be problematic in pristine environments, and both have to fulfill the needs of the receiving ecosystem.

Continuous plant operation

Irrespective of plant size, residual pretreatment chemicals in the RO brine blow-down are less hazardous. This major difference is due to dechlorination to protect polyamide membranes from oxidation, and chlorine has to be classified as a pollutant of high toxicity. Although residual chlorine levels are not only found in MSF plants but also in RO plants that operate on cellulose acetate membranes, polyamide membranes are increasingly used so that pressure on the marine environment will be further reduced in the future. If copper sulfate is added to control biofouling, treatment for membrane protection is not necessary. If copper is discharged, the major advantage of RO over MSF is diminished from an environmental point of view. However, this type of treatment does not seem to be very common. Further pretreatment chemicals are coagulants and scale control agents, which are additives in different drinking and surface water applications or sewage treatment. Although both are non-hazardous, some interaction with marine life may occur but can be prevented. For example, filter backwash can be deposited as solid waste to avoid increased turbidity or polyphosphate antiscalants can be replaced by new polymers to reduce eutrophication. Also of less concern is heavy metal discharge, although corrosion has been reported in RO plants. The high pressure piping is usually made from stainless-steel, so that iron and nickel ions as well as chromates and molybdates will predominate in the brine in concentrations that are not alarming.

On the contrary, corrosion of copper-based alloys in MSF plants has to be identified as a major source of brine contamination. The prevailing assumption that the brine copper content is diluted by cooling water to levels of no concern is not appropriate for three reasons. First, even the lowest reported discharge level is still slightly above reasonable water quality criteria for naturally low background levels. If corrosion control is not optimal, higher discharge concentrations will probably occur and have to be expected in many plants. Second, the cooling water will also be contaminated and finally, dilution is irrelevant if total loads are considered. For accumulating substances, concentrations in body tissue or sediments should additionally be measured to regulate and control the discharge. While removal from the water column reduces the area of impact for copper, dispersion of the second major pollutant from MSF, chlorine, is determined by hydrodynamic processes. Although self-decomposition is rapid, the remaining levels were found to be of toxicological importance even in relatively large distance from the plant. Furthermore, reactions with organic seawater constituents give rise to halogenated compounds and reduce the availability of free residual chlorine in the plant. To maintain the desired level, extra dosing becomes necessary which is not only a waste of material but also an unnecessary impact on the marine discharge site. The current practice of chlorination has a high potential for improvement. Through better process control, consumption can be reduced by up to 25 % (Shams El Din et al., 2000a) and this cost saving could probably compensate for the expense of treating the residual chlorine prior to discharge. This method should be further pursued as it takes environmental as well as economic aspects into account. With regard to copper, efforts are made to install improved construction materials in new plants whereas older plants become increasingly prone to corrosion despite continuous research in the field of material protection. In these cases, shut-down will eventually solve the problem when the plants are no longer operational or profitable, so that they have to be replaced by new facilities. As the other MSF pretreatment chemicals, i.e. polymer antiscalants and antifoamings, are relatively non-hazardous, emphasis is on copper and chlorine to make the MSF process more friendly to the marine environment.

Periodic cleaning

While pretreatment chemicals from the RO process are relatively non-hazardous, the opposite applies to membrane cleaning chemicals. This conclusion is based on the use of highly hazardous disinfectants (formaldehyde and isothiazole), detergents which interfere with cellular membranes until the active group has been split off (organo-sulfates and -sulfonates), and poorly degradable chelating agents that influence the environmental fate of heavy metals in the coastal water body (EDTA). As little is known on the current practice of waste disposal, discharge to the sea has to be taken into account. It can either be as direct blow-down immediately after cleaning or by storage and continuous blending into the waste stream. The first option has to be rejected as resulting concentrations will likely impair marine life until the chemicals are sufficiently diluted and degraded in the coastal water body. The alternative may prevent immediate effects from toxicity, but intentional dilution as a means of waste management must be questioned if pretreatment options are at hand, and is prohibited in many countries. Treatment includes deactivation of biocides by controlled reaction and primary degradation of detergents that render them relatively harmless (section 4.8). Problems of RO waste disposal are mostly due to the mentioned chemicals while pH of the solution is of minor concern. As seawater has a good buffering capacity, it will neutralize surplus hydroxide ions or protons almost immediately. Accordingly, the acidic cleaning waste from MSF plants had to be classified as relatively harmless, as it is only enriched with corrosion inhibitor at a level that is not alarming. However, cleaning volumes are considerably larger and precautionary measures should not only be considered if water exchange is restricted. Neutralization prior to disposal and relocated, improved outfall structures are alternative options.

5

The Arabian Gulf

"Realizing that pollution of the marine environment in the region shared by Bahrain, Iran, Iraq, Kuwait, Oman, Qatar, Saudi Arabia and the United Arab Emirates, by oil and other harmful or noxious materials arising from human activities on land or at sea, especially through indiscriminate and uncontrolled discharge of these substances, presents a growing threat to marine life [...]
Mindful of the special hydrographic and ecological characteristics of the marine environment of the region and its particular vulnerability to pollution [...]
The contracting states shall take all appropriate measures to prevent, abate and combat pollution caused by discharges from land reaching the sea area whether water-borne, air-borne, or directly from the coast including outfalls and pipelines."

Kuwait Regional Convention (1978)

5.1 Pollution of the Gulf

The Arabian Gulf is a part of the ROPME sea area, which is the Regional Organization for the Protection of the Marine Environment, and comprises the coastal waters of the above mentioned member states. This marine environment has always been a valuable resource that supported the coastal population, and it still does in a broader sense, as the current living standard is based on oil exploitation and desalination of seawater. With regard to these two "strategic fluids" (UNEP, 1999), the Arabian Gulf remains unsurpassed: It contains 46 % of the world's proven oil reserves (Subba Rao and Al-Yamani, 2000) and about half the world's seawater desalination capacity is installed in the region, which is approximately 7 million m^3 per day compared to a total of 15 million m^3 (based on Wangnick, 1999). The high share in desalination can clearly be attributed to the extensive oil reserves of the Gulf countries, which allow for a relatively cheap production in large thermal plants. On the other hand, desalination and oil exploitation are competing uses of the marine environment. The Arabian Gulf is one of the most heavily polluted water bodies with regard to petroleum components[a]. Of these, volatile compounds can be chlorinated during the desalination process and co-distill with the water, thereby contaminating the product (Ali and Riley, 1989; Saeed et al., 1999), while oil slick might enter the plant along with the feedwater. This will irreversibly damage RO membranes or require extensive cleaning in MSF plants. Oil contamination is not the only concern for drinking water production. Water usually becomes scarce when major industrial complexes are being constructed or the local population increases, and coastal areas in the Gulf presently experience a tremendous growth (Abdel-Moati and Kureishy, 1997). To satisfy the demand, new desalination plants arise where at the same time seawater is polluted as a result of this development. Effluents are generally dumped to the Gulf without proper treatment (Hinrichsen, 1998), and little has been done to protect the marine environment from land-based activities (UNEP, 1999), since the Gulf member states have agreed upon combating pollution in 1978.

However, in fulfillment of the Kuwait Convention and recognizing that quality of life is closely connected to an intact Gulf ecosystem, a more recent protocol entered into force in 1993[b]. Furthermore, each gulf country has meanwhile implemented its own marine environmental laws (Subba Rao and Al-Yamani, 2000), which for example establish procedures for environmental impact assessment in the planning stage of new projects or effluent guidelines for disposal of liquid waste. It is evident that sustainable

[a] This is caused by 800 offshore platforms, 25,000 tankers transporting oil through the Gulf per year (Hinrichsen, 1998), periodic tanker accidents, the 1991 Gulf War oil spills and effluents from petrochemical industries (Saeed et al., 1999).

[b] The Protocol for Protection of the Marine Environment against Pollution from land-based sources was signed in February 1990 by the mentioned contracting states and entered into force in January 1993 (UNEP, 1999).

use of seawater requires regulations for all sorts of waste discharge to the sea, not at least to guarantee a constant quality of seawater for drinking water production. As power and desalination plant effluents are themselves a major source of pollution in the Gulf (UNEP, 1999), the same standards have to apply to these facilities.

In this chapter, attention is therefore given to pollution of the Arabian Gulf through seawater desalination plants. As copper and chlorine were identified as substances of major concern, both will be considered in terms of total loads. In addition, antiscalants are of interest with regard to eutrophication, if polyphosphates are used and emphasis is also given to new polymers, of which little is known regarding their environmental fate and effect on the marine environment.

5.2 Copper loads

Copper concentrations in the desalination discharge and Gulf seawater were discussed in section 4.7. Supplementary, total copper loads will be considered in the following to take into account that most of this material will remain in the Gulf, where it adsorbs to suspended material and sinks to the seabed.

Assuming again 15 μg/l brine concentration for a plant the size of Doha West with a brine volume of about 0.8 million m^3/day, the copper load would amount to approximately 12 kg/day. With regard to the unknown contribution of MSF and power plant cooling water, and the potential for effluent variability due to acid cleaning or local pitting and corrosion, 15 μg/l and 12 kg must be understood as conservative estimates. The actual discharge will probably be higher, i.e. brine concentrations of less than 100 μg/l would result in 7-fold higher copper loads, as achieved in "modern plants with good process control" (Oldfield and Todd, 1996, Table 4.1). Projected to the whole Gulf, the daily copper input through MSF plants could range between 200-1330 kg, with annual discharge loads of 73-485 t (Fig. A.4 on page 113). These loads must be compared to estimates of natural copper loads: the Arabian Gulf has a volume of approximately 6000 km^3 (Anderson and Dyrssen, 1994) and if a concentration of 0.6 μg/l[c] is assumed to be close to ambient, the whole sea would con-

[c] Original value of 9.8 nM/kg, based on Abdel-Moati and Kureishy (1997).

tain 3,600 t copper. The input through desalination consequently amounts to 2-13 % of the natural copper load of the Gulf per year. However, it has to be expected that this extra material is not dispersed uniformly in water and sediment. Large-scale circulation will have little influence on spreading, although a drift of material may be perceptible locally before copper is deposited in sediments. Concern consequently arises for areas of restricted water exchange and high installed capacities, like the Gulf of Salwah (between Qatar and Saudi Arabia), the barrier island complex of the UAE and Kuwait Bay, where sedimentation might amount to unacceptable copper loads.

Although several studies investigated copper contamination in Gulf sediments, only one pointed out that desalination activities might contribute to increased copper levels but could not detect a clear signal for this hypothesis: Copper levels along the Saudi Arabian coastline were comparable to elsewhere in the Gulf, but were correlated to organic-rich, fine-grained particles and therefore higher in seagrass beds (2.1-4.3 mg copper/kg sediment, Coles and McCain, 1990). Similarly, mangrove stands are prone to copper accumulation (5.3-29 mg/kg, UAE, Shriadah, 1999). Both ecosystems are areas of high primary production, in which organic material is abundant. Thereby, dissolved copper can be complexed by humic substances, adsorbed to suspended matter and deposited in calm water that allows for a long enough sinking time, e.g. among mangrove roots. On a larger scale, a similar trend was observed by Basaham and Al-Lihaibi (1993) in the northwestern Gulf. Copper decreased from 34 mg/kg in Kuwait, where muddy sediments predominate due to the Shatt Al-Arab estuary, to 1 mg/kg in relatively coarse-grained sand sediments around Bahrain and Qatar. This observation correlates with a high copper content in Kuwaiti coastal water: 25 μg Cu^{2+} per liter in the intake water of Doha (Bou-Hamad et al., 1997) considerably exceeds seawater levels found around Qatar of 0.6-3.4 μg/l (Abdel-Moati and Kureishy, 1997). As fine silt and clay are predominant in Kuwait Bay sediments, they are also suspended in the water column and have serious effects on desalination plants in Kuwait by settling in intake channels and heat exchanger tubes (Al-Awadi, 1994). The high copper content of this material can either be explained by continuous pollution and accumulation, or it is already present in the load transported by the Shatt Al-Arab. As Basaham and El-Sayed (1998) observed similar copper to aluminum ratios in sediments of the Gulf and those of the Shatt Al-Arab estuary (Gulf 17.2:1, Euphrates 17.4:1, Tigris 10:1), it can be concluded that the riverine input will have a major influence on copper levels in the Gulf.

However, the comparatively high copper levels in the northern Gulf may be further increased through several sources of contamination. Besides desalination and industrial or urban effluents, emphasis is given to the input caused by oil spills, especially during the 1991 Gulf War. Although oil-related metals are predominantly vanadium, nickel, chromium and lead (Fowler et al., 1993), copper and other trace metals are usually included in surveys of oil pollution (Kureishy, 1993; Sadiq, 1996). These copper signals will probably overlap with input from desalination, but overall, copper levels are still relatively low even in heavily oil-contaminated sites. In Saudi Arabia, which was badly affected by the 1991 oil spills, levels ranged between 3.14-5.53 mg copper/kg sediment (Fowler et al., 1993) which has to be compared to other impacted areas of the world, e.g. 7-2398 mg/kg in English estuaries or 0.06-218 mg/kg in US estuaries (Kennish, 1997).

Effects from copper contamination in the Arabian Gulf can hardly be predicted. Attention should be given to the affected sub-ecosystems, as mangroves and seagrass beds have a high diversity and productivity and are therefore important for the whole food web of the Arabian Gulf. Questions of copper toxicity and accumulation must be raised again, but depend on metal speciation. Metals incorporated in the lattice structure of minerals are non-bioavailable, similar to metals of anthropogenic origin, which have been adsorbed, complexed and deposited in sediments (Kersten and Kröncke, 1990). However, re-suspension and -dissolution cannot be excluded in shallow coastal areas, where tidal currents are strong and sediments are turned over by benthic organisms. Although dissolved copper is more bioavailable, some copper complexes may be metabolized and furthermore, many benthic invertebrates feed on suspended or deposited material by digesting its organic content. Among these organisms, very high bioconcentration factors[d] have been observed for copper, e.g. up to 30,000 for molluscs (HSDB 2001r). Copper enrichment in organisms therefore reflects the bioavailable fraction of copper in the environment. The

[d] The ratio of the element's concentrations in the aquatic organism divided by its concentration in water.

use of bioindicators to detect pollution is restricted by the fact that not all species provide reliable information, and suitable organisms might not be present at the site of investigation. To compound matters, data on sex, stage of development, etc. influence metal accumulation. With regard to these problems, single studies will be appropriate to monitor health limits in edible species, but will not be appropriate to detect spatial and temporal trends of pollution. Biomonitoring in the Gulf requires combined action, as indicators have to be identified that are ubiquitous in the Gulf and suitable for the specific element. The 18-month marine monitoring programme co-ordinated in the ROPME states (UNEP, 1999) may be a promising start, and results have to be awaited.

To conclude, clear signals of copper contamination in water, sediment and organisms attributed to desalination activity are missing, but the calculated amounts indicate that desalination plants require high attention. This is so far given to copper input through oil pollution, sewage and other industrial effluents, which will probably overlap with heavy metal input by desalination plants in urbanized areas. Although copper contamination is still relatively low in the Gulf, total copper discharge has to be restricted, especially as important sub-ecosystems are prone to copper accumulation. Special concern arises for the Arabian Gulf as a whole, which is a sedimentary basin dominated by soft substrate (Sheppard et al., 1992), and probably a sink for copper.

5.3 Chlorine loads

In contrast to copper, environmental fate of chlorine is dominated by dispersion and rapid degradation, but major environmental impacts have still to be expected by low residual levels in coastal waters (cf. section 4.3.1). In addition, large discharge volumes of positively buoyant cooling water and desalination brine will extend over large coastal areas, and to compound matters, total chlorine loads are often increased by power plants close by. Calculations have to include both, electricity and water production, as they are usually coupled and the desalination plant receives its energy and steam from the power plant.

For example, the Umm Al Nar power and desalination plant in the UAE chlorinates approximately 10 million m^3 per day (Shams El Din and

Mohammed, 1998), and is similar in size to the Doha West plant in Kuwait with a daily seawater intake of about 10.5 million m^3 (cf. Table 4.1). More than half of this volume is used as cooling water for the electricity generating turbines, while the rest serves as feedwater for the desalination process. To control biological growth inside the plants, both feed streams are treated with an average of 2.5 mg/l chlorine at the site of Doha, which amounts to a total daily consumption of 26,000 kg chlorine. However, most of this material is degraded in the plant, and it can be assumed that only about one tenth of the initial concentration is discharged, i.e. about 250 μg/l residual chlorine. Effluent levels may be higher in Doha with a maximum of 500 μg/l specified by Abdel-Jawad and Al-Tabtabaei (1999), so that a total of 2,500-5,000 kg could enter the coastal waters near Doha each day.

Further self-decomposition of chlorine in warm seawater and direct sunlight will account for a rapid decrease of environmental concentrations, but low values may still be toxic to marine life. As discussed in section 4.3.1, estimated residual levels of 50 μg/l still exceed an acute EPA WQC of 13 μg/l by a factor of four in 1 km distance from the outlet. For comparison, daily waste water volumes of Doha West would alone fill a circular area with a radius of about 1200 m and 2 m depth as illustrated in Fig. A.7. The outfall site of Doha West is located in the intertidal of Sulaibekhat Bay, and considerable plume spreading has to be expected as most areas are shallower than 2 m with a maximum depth of 5 m where the inlet opens into Kuwait Bay. Location of outfall structures in shallow coastal areas should be avoided and may be prevented by environmental regulations, e.g. the US EPA limits the toxic dilution zone to five times the local water depth. In the case of Doha, this zone had to be limited to 10 m distance at most for meeting the acute WQC of 13 μg/l. It is evident that this restriction cannot apply to large discharge rates of more than 100 m^3/s as in the case of Doha. But alternatively to extending the impact area, neutralization can be considered to achieve levels that will not be lethal to marine organisms. It may also be objected that water of the Bay is flushed out effectively once a day. Kuwait has a diurnal tidal pattern with high tidal ranges between 2-4 m (Sheppard et al., 1992), and tides are important factors for mixing waters and dispersing pollutants on a small scale of less than 10 km (Reynolds, 1993). On the other hand, the incoming tide will restrict the flow out of the Bay, so that pollutants can accumulate before they are flushed out again.

Impacts will also be more severe in the intertidal during low tide, when the discharge cannot be properly diluted by seawater.

Pollution by effluents with high concentrations of chlorine has been reported for Sulaibekhat and Dawhat Kazima mudflats in Kuwait Bay, which are highly productive ecosystems and sustain numerous migratory birds as well as commercial fish and shrimp species (Scott, 1995). By the current practice, acute concentrations for marine zoo- and phytoplankton will very likely be exceeded in the coastal waters. These organisms are ubiquitous and have short generation times, so that overall population dynamics will not suffer from chlorination. On the other hand, the area which is permanently exposed to residual levels of this biocide is large. Drifting larvae of benthic organisms and fish could also be affected when they pass through the plume and their survival is prerequisite for new recruitment in the discharge site. Furthermore, acutely toxic effects on adults have to be expected if they are not able to avoid the area of highest concentration, while chronic levels will still occur in a far larger distance than 1 km from the plant. This assumption is based on the estimated extensions of the mixing zone.

With regard to the whole Gulf, chlorination may amount to 15,000 kg per day (about 5500 t/a) as shown in Fig. A.5 on page 114, which is based on estimated residual levels of 250 μg/l in desalination cooling water and brine discharge. It has to be expected, that at least the same amount is added to the Gulf by power plants, although installed capacities could not be obtained. Furthermore, discharge levels may be higher, as levels are usually established by trial and error and depend on water quality (Shams El Din et al., 2000a), or during shock treatment of the plant. It is unknown if this total load will have adverse effects on the whole Gulf ecosystem, but impacts are probably more restricted to local discharge sites. Although self-degradation of chlorine occurs rapidly, chlorinated organic compounds can be rather persistent. These substances are more likely to accumulate in the counterclockwise current in addition to other pretreatment chemicals, thereby deteriorating the water quality not only for marine life but also for the downstream desalination plants (Hoepner, 1999; Höpner and Windelberg, 1996). Further attention should be given to areas of restricted water exchange, e.g. shallow coastal inlets in Kuwait Bay but also to the Gulf of Salwah, the coastal waters of Bahrain and the UAE. Some of their coral reefs, mangrove swamps, seagrass beds, shallow low-energy bays and semi-enclosed lagoons are under threat from human activities (Scott, 1995), among which are desalination plant effluents. On the other hand, these sub-ecosystems were ranked highest on a sensitivity scale with regard to emissions from these plants (Höpner and Windelberg, 1996).

5.4 Antiscalant loads

Most parts of the Arabian Gulf lie within the photic zone, so that its benthic communities contribute a high share to the total production of the Gulf, while pelagic productivity is comparatively low (Sheppard, 1993). Phytoplankton growth depends on the availability of essential macronutrients and trace metals dissolved in seawater, e.g. phosphates or iron. As the Arabian Gulf must be classified as an oligotrophic sea, drastic effects have to be expected if nutrient levels are increased through human activities (Hoepner, 1999). Effects of eutrophication have been observed in many other polluted seas, and similar symptoms like coastal algal blooms, red tides and higher jellyfish abundance have increasingly occurred in the Gulf, too, and might be due to large-scale eutrophication (Price, 1993). In this context, polyphosphate and polymer antiscalant discharge from desalination plants have to be considered as they might interfere with nutrient balances in the Gulf.

Polyphosphates

As discussed in section 3.3.2, polyphosphate dosing is still a common pretreatment in RO plants and alternatively, organophosphate antiscalants are often used (an overview on scale control methods in commercial plants is given in Al-Shammiri et al., 2000). Although enhanced primary production could clearly be attributed to phosphate discharge locally, e.g. close to the outfalls, effects of this practice are difficult to assess in the coastal water body as a whole. Desalination plants are usually located where potable water is required, so that urban and industrial areas will add surplus nutrients through several point- and non-point sources. Increased natural background levels aggravate the detection of relative effects from desalination activities. To illustrate the waste load, it will therefore be compared to natural sources of phosphate input like riverine transport. The Shatt Al-Arab, which is the main freshwater source in the north, causes a north-south gradient in essential macronutrients, followed by a similar trend in primary production (Subba Rao and Al-Yamani, 2000). The annual runoff decreased substantially over the last twenty years, as its water is diverted for agricultural and industrial purposes, but more recently estimated to be 45 km^3/a (Reynolds, 1993). Its nutrient load is thought to be increased by human activity but current values could not be obtained. With regard to phosphate ranges between 1.55-6.01 μmol/l measured in 1985 (Subba Rao and Al-Yamani (2000) after Saad), the riverine transport might amount to 6,000-26,000 tonne/year. For comparison, the total discharge from desalination would be 800 tonne/year, if all RO plants were treated with 2 mg/l polyphosphates[e] (Fig. A.6). However, this estimate is based on a comparatively low polyphosphate dosage, which could be five times higher (cf. Table 3.5 on page 34).

With regard to these findings, it must be concluded that discharges from desalination plants will have minor effects on the whole nutrient balance of the Gulf, but local impacts may be significant. For example, it was calculated in section 4.5.1 that large RO plants can consume about 200 t/a and 60 % of the installed capacity in the Gulf is achieved in two large plants: Az Zor South and Al Jubail in the northwest (Fig. A.3 on page 112). As little is known on their pretreatment methods, polyphosphate addition cannot be excluded. In this case, stringent discharge criteria could be implemented to reduce phosphate input, and point sources are far more easy to control than diffuse drainage. For example, a limit of 0.025 mg/l total phosphorus was suggested for RO waste disposal into a pristine water body with naturally low nutrient levels (EPA, 1999). Alternatively, polymaleic and polycarboxylic based antiscalants can be used. The shift from conventional polyphosphates to these new polymers in MSF plants already helps to restrict nutrient discharge, as their estimated daily input to the Gulf is considerably higher due to the predominance of thermal plants in the region.

[e] A dosage of 2 mg/l SHMP $(NaPO_3)_6$ corresponds to 1.88 mg/l PO_4-P following complete hydrolysis ($(NaPO_3)_6 + 6\ H_2O \rightarrow 6\ Na^+ + 12\ H^+ + 6\ PO_4^{3-}$) and equals 0.6 mg/l total phosphorus.

Polymer antiscalants

Discharge concentrations of polymer antiscalants are rarely published but Morton et al. (1996) specified a residual of about 0.5 mg/l polymer additive in the effluent of an MSF plant, that was dosed at 2.25 mg/l. The effluent level can be put down to process design, i.e. the brine is blended with a threefold amount of cooling water from the heat rejection section, which is not treated with antiscalants. Therefore, most of the added polymer will be discharged to the sea, although break-down by high temperature or reaction with chlorine might diminish total loads. Based on a typical dosage of 2 mg/l for polymer antiscalants, pretreatment in a large MSF plant like Doha West will require about 2500 kg per day (Table 4.1). Projected to all MSF plants in the Gulf, the daily input is estimated to 40,000 kg/day (14,600 t/a, Fig. A.6). Discharge levels from the desalination plant will be further reduced if the waste stream is blended with cooling water from power plants, e.g. 0.2 mg/l residual polymers will be present in the effluent following tenfold dilution. Further dispersion occurs in the coastal water body so that environmental levels will be relatively low, and immediate effects of toxicity can largely be excluded. On the other hand, immense loads enter the Gulf each day which have a half-life of roughly one month or more, i.e. are comparatively slowly degraded. Little is known about the environmental properties of polymer antiscalants, but it has to be expected that they behave similar to humic substances (cf. section 4.5.2). Both complex metal ions in seawater, and antiscalants are used in desalination plants because of this property. Therefore, the influence of additional complexing agents will depend on natural background levels of humic material. This is alarming as the Gulf contains only minor amounts of these natural seawater constituents, except for estuarine, mangrove and intertidal mudflats, where organic and humic compounds are generally more abundant. With regard to the pelagic water body, marine primary production and riverine transport of terrestrial material account for most of the input (GKSS, 1986), but the Gulf is an oligotrophic sea with low freshwater run-off. Further attention should therefore be given to the influence of antiscalants on metal solubility, mobility and availability. For example, it is of interest if complex formation delays metal deposition in sediments or even enhances re-dissolution and dispersion within the ring-shaped current of the Gulf. As complexes are generally non-bioavailable, discharge of antiscalants might also reduce toxicity of some metal ions or withdraw essential trace nutrients from the water column. Although new polymers help to reduce the overall phosphate input to the Arabian Gulf and are non-hazardous at a first glance, it must be concluded that their environmental fate is poorly known. This situation would be less critical, if total discharge loads were lower. With regard to the growing desalination industry in the region, however, the risk of adverse effects cannot be excluded on a longer time scale.

6

Summary and Outlook

Two common assumptions in the discussion of desalination waste have to be evaluated in the context of this work, and will also help to summarize the presented results. The first is *that the process predominantly concentrates what is already in the seawater, thereby the waste constituents are mostly substances of natural origin to which marine organisms are adapted.* It has been shown that strong deviations in salinity from ambient levels may also be classified as pollution (cf. section 4.2.1). Organisms normally live within their optimal range, while de- or increased values may be at the rim of their tolerable limit. However, direct effects of brine disposal can be avoided, either by dilution with cooling water of ambient salinity (usually MSF plants) or other waste streams in the case of single purpose RO plants. As minor deviations are usually of less concern, this method for effect mitigation is acceptable, as it only reverses the process of concentrating seawater. Therefore, one can agree upon the above argument with regard to natural seawater constituents, if plant design provides for adequate blending, but dilution of residual chemicals is generally no means of waste water disposal and often prohibited.

This leads up to the second prevailing opinion, *that discharge concentrations of pretreatment chemicals are already low, and that they are further reduced by dilution with cooling water.* The necessity of waste stream blending to reduce salinity seems to solve the problem of residual chemicals at the same time. The argument applies to antiscalants and antifoamings, which are added to the plant's feedwater but not to the cooling water, so that dilution will effectively reduce overall concentrations prior to disposal. However, both substances are of minor concern with regard to their toxicological potential at low concentrations (cf. sections 4.5.2 & 4.6), and even the brine could be discharged directly if it was only for these chemicals. Dilution unfortunately focuses on brine constituents that are not central to the impact assessment of desalination waste, while the discussion is not appropriate for substances of importance. Chlorine concentrations are about the same in cooling water discharge and brine, so that total loads are even increased (section 4.3.1). The same argument applies to temperature, which is elevated in both waste streams (section 4.2.2). This physical parameter may influence marine life in the mixing zone similar to modified salinity values, but expected effects are of minor importance if compared to chemicals of high toxicological potential. In the case of copper, waste stream blending may reduce total concentration in the discharge, as contamination is expected to be lower in the cooling water. However, this procedure will still increase total copper loads (section 4.7). Total amounts are not only of concern for accumulating substances like copper, but for every substance that is added to the feedstream. Although concentrations are low, the daily and annual cumulative discharge is immense, as discussed for single chemicals in chapter 4. Special attention is therefore required for areas of high installed capacity like the Arabian Gulf (chapter 5).

To conclude, the effluent should be treated in the best practicable manner. This particularly applies to membrane cleaning solutions and the continuous discharge of MSF plants (cf. section 4.9). Treatment of RO cleaning waste has the advantage that total volumes are low, and chemicals can be deactivated or degraded prior to discharge. In contrast, flow rates are high during normal operation of MSF plants and posttreatment of e.g. 10 million m^3/day seems to be impossible. However, this predominantly requires dechlorination, and it can be calculated that extra expenses will not cause a significant rise in product costs (section 4.3.1). No options are currently at hand for copper treatment besides optimization of process control to reduce corrosion. In addition, new construction materials are investigated and might solve the problem of copper discharge in the future. Research for better plant performance often reduces environmental impacts at the same time. For example, an optimized chemical dosage also reduces residual levels in the brine, or new compounds like polymer antiscalants reduce eutrophication and heavy metal discharge by displacing phosphate treatment and limiting corrosion. Environmental impact mitigation, however, should not be a mere side effect but an integral part of new developments in desalination technology.

Outlook

In this context, little is still known about polymer antiscalants, while at the same time high amounts are released into the Gulf each day (cf. section 5.4), or about new tubing materials like titanium or plastic, which might replace copper alloys in the future. Prerequisite for an analysis of environmental fate and impact is a better general knowledge of the discharge. Information on brine composition and residual concentrations is incomplete, and to compound matters, treatment methods and chemical dosage vary widely among single plants.

In addition to effluent properties, more field investigations are required and should include water and sediment sampling as well as bioaccumulative studies with regard to heavy metals, especially copper. Other substances, e.g. halogenated organic compounds or polymer antiscalants, have a high mobility and persistence in the marine environment and should be considered in the context of water circulation. Effluent modeling (cf. last paragraph) will therefore not only provide valuable information on plume spreading but also on chemical dispersal.

Field observations can be accompanied by laboratory tests on toxicity. Toxicological results were consulted for a comprehensive number of conventional and toxic pollutants in the desalination discharge so far (chapter 4), but information is still incomplete for some substances. Little is also known about additive or synergistic (more than additive) effects on marine life, although most single pretreatment chemicals are relatively non-hazardous. To take the presence of more than one substance into account, discharge regulations might ask for a reduction of each acceptable limit (GESAMP, 1986), but how can reasonable levels be accomplished ? The underlying mechanisms of toxicity are not easily revealed, which is often difficult enough for one-cause one-effect relationships. To compound matters, test species are not equally sensitive and toxic concentrations also depend on health or reproductive state of single organisms. Laboratory results, which are usually established on freshwater species, are therefore not necessarily representative of the real situation although they provide a good overview on toxicity. To compensate for the insufficient knowledge of multi-component wastes and their burden on the marine environment, desalination-specific tests could be conducted. By exposing preferably local saltwater species to the whole effluent, experimental design can account for additive and synergistic effects of single components. A similar procedure has proven effective by uniting information from single substances on different marine species in the form of quality criteria for seawater. As these standards provide valuable information for the assessment of effluents, the same has to be expected for whole effluent tests. In addition, this approach takes conventional parameters into account. Organisms exposed to high salinity, temperature and turbidity or low pH and dissolved oxygen levels might exist close to their tolerable limit. The resulting stress can impair their resistance to residual chemicals. For desalination plants in the USA, this proof of non-toxicity has become obligatory to receive a discharge permit for surface waters (Mickley, 1995).

To further improve knowledge on specific discharge situations, plume modeling is required. Discussion in this work was restricted to general aspects of salinity, temperature and density as local conditions will differ among plants. These include topographical and hydrographic parameters in addition to detailed information on the practice of discharge. For example, some parameters of importance are water and discharge depth, ambient density including seasonal variations, wind- and density-driven currents, tidal regime, diameter of the outfall, effluent velocity and volumes etc. Modeling is recommended in the permitting process of industrial, municipal, thermal and other point discharges into coastal waters by the US EPA (Jirka et al., 1996) and the EPA Cormix model is also applicable to desalination plant discharge (Del Bene et al., 1994). This tool helps to determine the best outfall location by acting out different scenarios and identifying the ecosystems in reach. For example, positively buoyant effluents may be best injected into deep water to increase mixing of the rising plume, or multiport diffusers may account for the same effect. Furthermore, offshore disposal will avoid shore-hugging plumes that affect sensitive coastal habitats, which are already under stress from several sources of pollution and general coastal development.

A

Maps of the Arabian Gulf and Red Sea

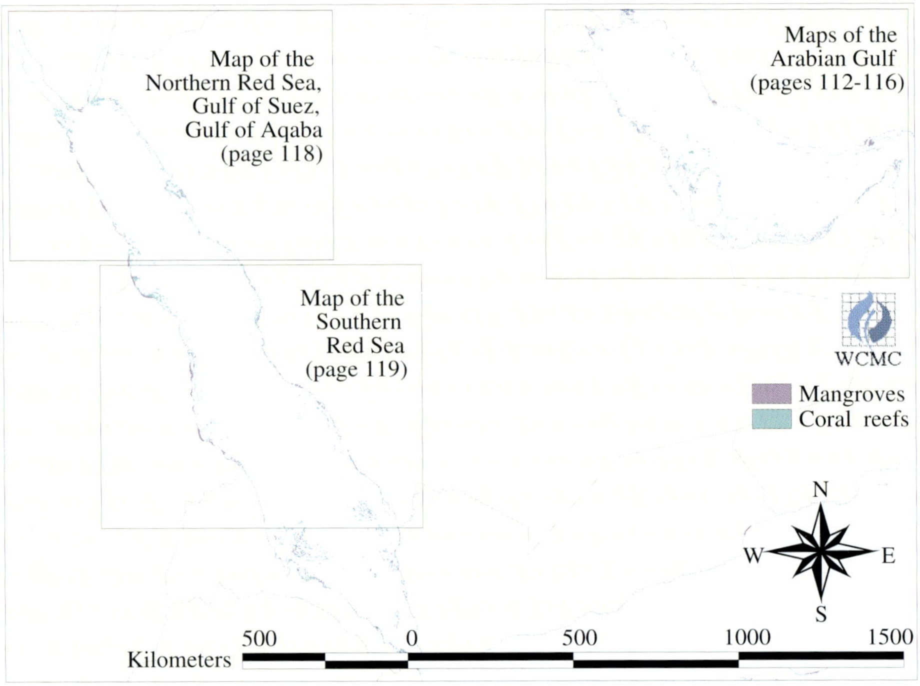

FIGURE A.1: Overview map of the Red Sea and Arabian Gulf

Figures and maps of the Arabian Gulf in the Appendix:

- Capacities by country, Fig. A.2
- Plant locations and capacities, Fig. A.3
- Daily copper input by MSF plants and water circulation, Fig. A.4
- Daily chlorine input by MSF Plants, Fig. A.5
- Daily antiscalant input by MSF and RO plants, Fig. A.6
- Plants in the Bay of Kuwait, Fig. A.7

Figures and maps of the Red Sea in the Appendix:

- Capacities by country, Fig. A.2
- Plant locations in the northern Red Sea, Fig. A.8
- Plant locations in the southern Red Sea, Fig. A.9

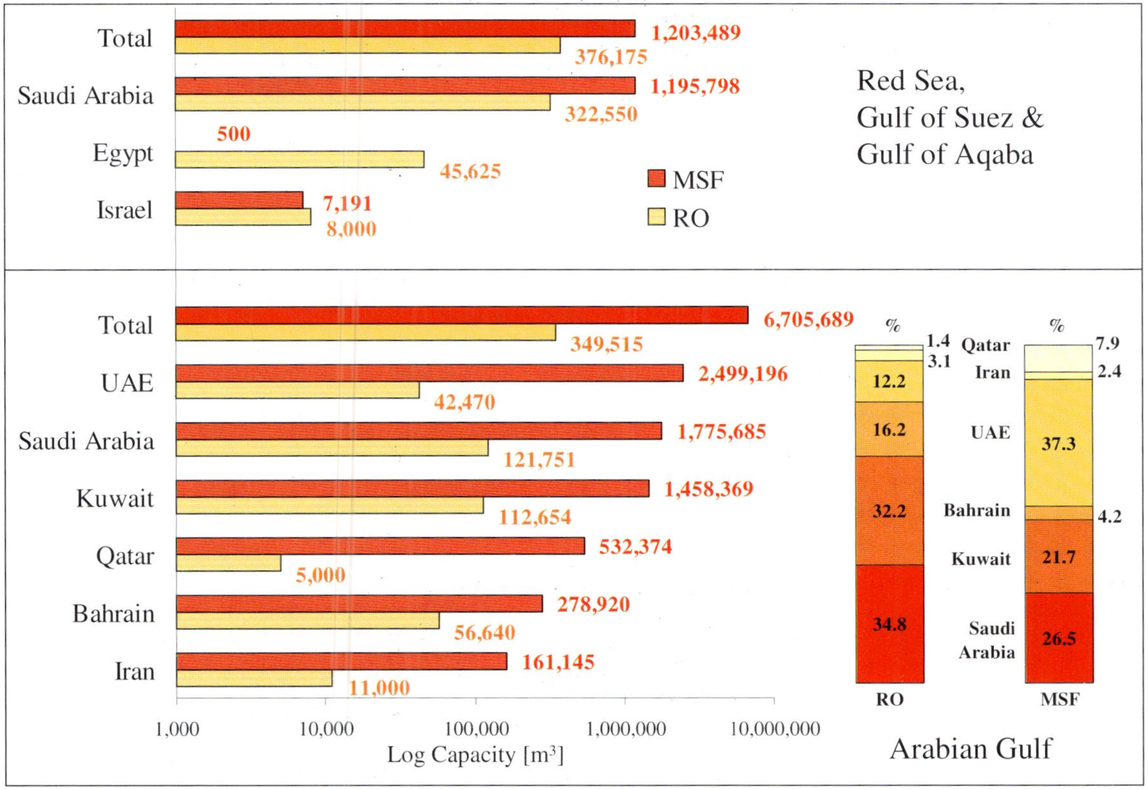

FIGURE A.2: MSF and RO capacities in the Arabian Gulf and Red Sea, including all plants shown in Fig. A.3, A.8 and A.9 that operate on seawater. Data source: Wangnick (1999). While total RO capacities are about equal in bcth regions, the installed MSF capacity is more than five times higher in the Arabian Gulf.

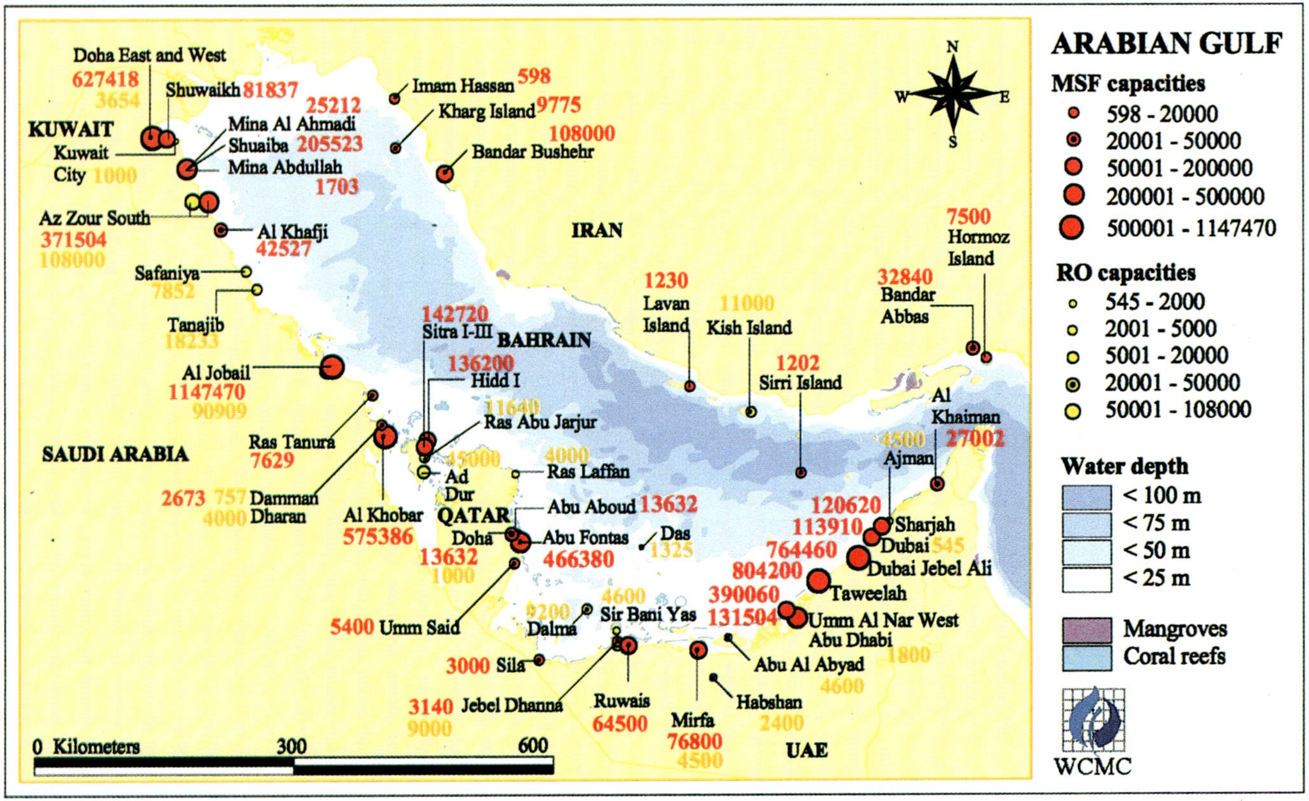

FIGURE A.3: MSF and RO plants in the Arabian Gulf. Data on coastline, corals and mangrove areas supplied by and copyright to World Conservation Monitoring Centre (WCMC) (2001). Included are all MSF and RO seawater desalination plants, of which locations could be identified. These account for more than 95 % of the overall capacity in the Arabian Gulf, based on Wangnick (1999). Water depth based on Gita Shenasi Cartographic & Geographic Institute.

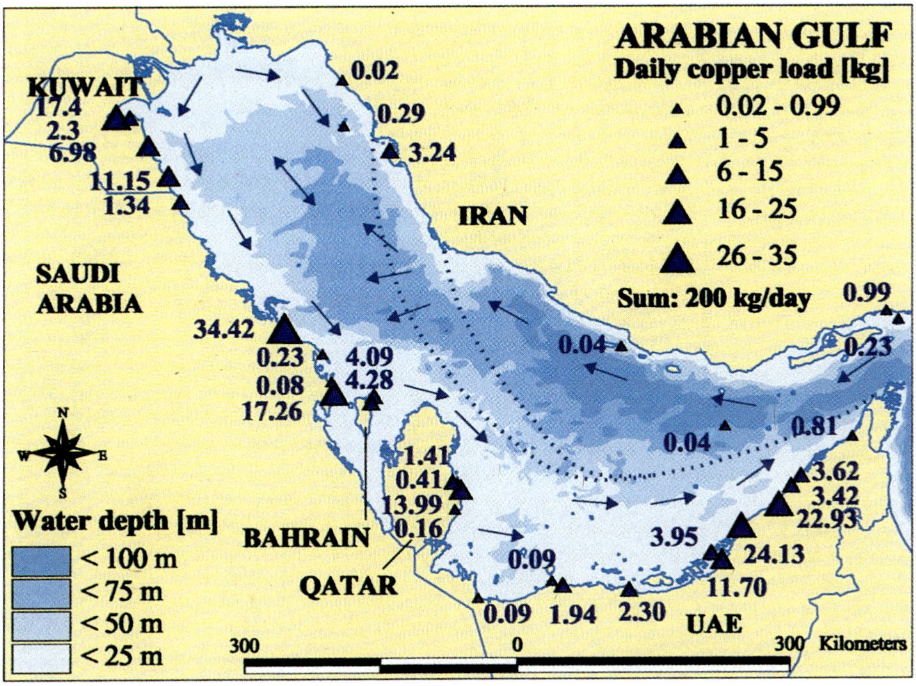

FIGURE A.4: Estimated daily copper input to the Arabian Gulf by MSF desalination plants, based on a minimum of 15 μg/l copper in the brine blow-down. Not included is copper contamination in the cooling water from the reject section of the MSF plant as well as discharge from adjacent power plants, that will further increase total loads. The annual discharge through desalination (73 t) therefore adds 2 % of the natural copper load to the Gulf. This estimate must be understood as conservative and contamination is probably higher (cf. section 5.2 on page 95). As copper mobility is reduced by transfer from dissolved copper to suspended material, the shown water circulation (Reynolds, 1993) will influence copper dispersion only until it is deposited in sediments. The main Gulf circulation is density-driven: high evaporation causes sinking of highly saline water and formation of a bottom current which flows out of the Strait of Hormuz. Surface inflow from the Arabian Sea compensates for this water loss and occurs along the Iranian coast. Dotted lines mark the approximate limit of this inflow, which extends further north in summer and is forced back by a wind-driven surface current in winter. These south-easterly Shamal winds also cause a corresponding water movement along the Saudi Arabian shoreline, which stagnates east of Qatar, where high evaporation and sinking predominates.

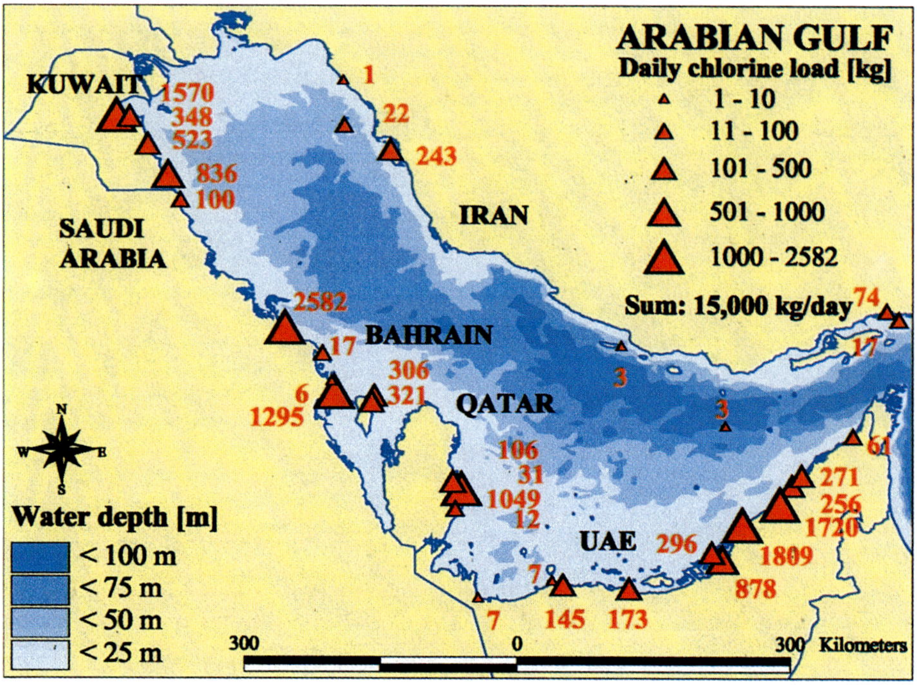

FIGURE A.5: Estimated daily chlorine input to the Arabian Gulf by MSF desalination plants, based on residual chlorine levels of 250 $\mu g/l$. As the desalination plant's cooling water is also chlorinated, it is included in the estimate. Not quantified are chlorine discharges from power plants, which are often co-located to MSF plants and increase the total load to the Gulf considerably. See also section 5.3 on page 98.

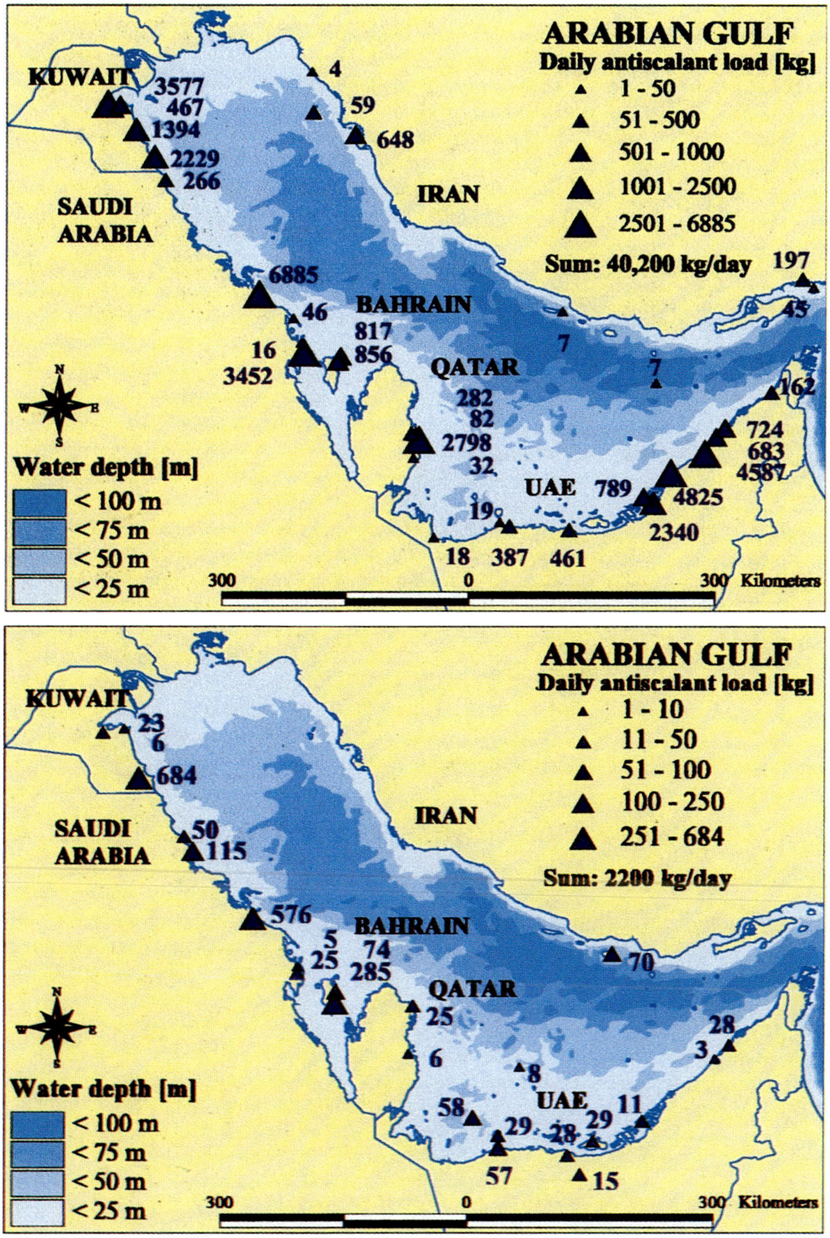

FIGURE A.6: Estimated daily antiscalant input to the Arabian Gulf by MSF plants (top) and RO plants (bottom), based on an average dosage of 2 mg/l to the desalination plant feedwater. While MSF plants predominantly rely on polymaleic, polyacrylic and phosphonate antiscalants, RO plants may also operate on polyphosphates (cf. section 5.4 on page 101).

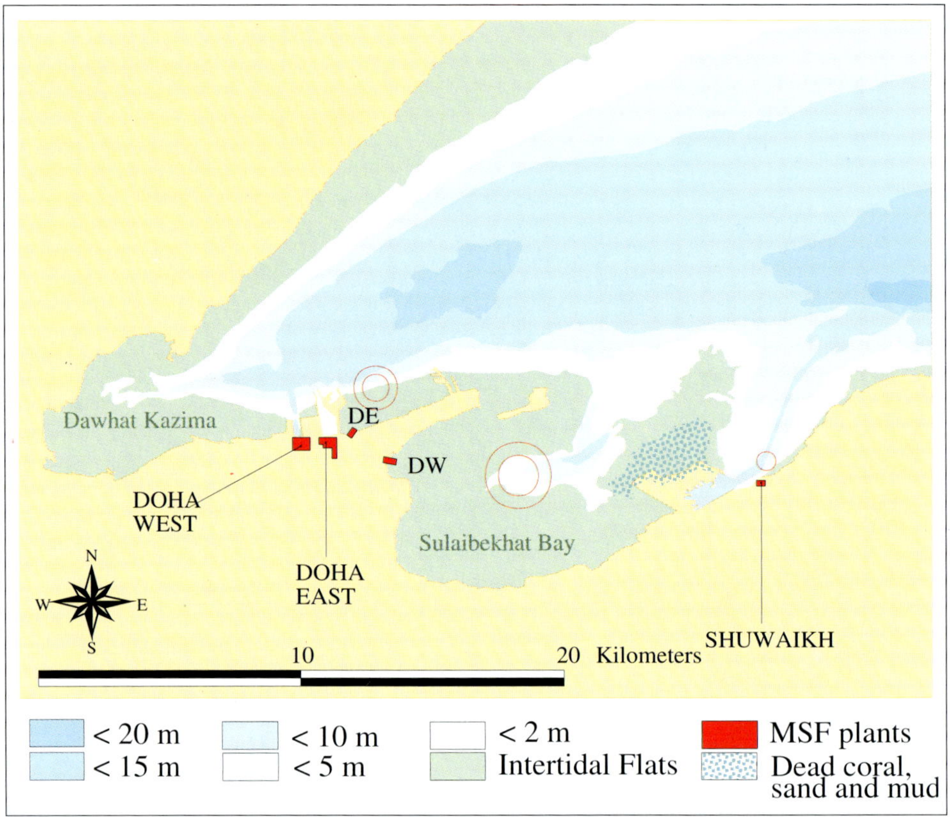

FIGURE A.7: The Bay of Kuwait. Water depth and plant locations based on UK Hydrographic Office (1990) and Saeed et al. (1999). DE: discharge site of Doha East, DW: discharge site of Doha West. The red circles illustrate volumes of daily desalination discharge (inner circle) and total effluents (outer circle), assuming a water depth of 2 m. For example, the outer circle in Sulaibekhat Bay covers an area of about 500 ha (radius of 1260 m), which is roughly 20 % of the intertidal mudflats of the Bay (2250 ha, Scott, 1995). The main plant properties of Doha West, Doha East and Shuwaikh are summarized in Table A.1 on the facing page, which also contains estimated chemical discharges.

TABLE A.1: Plant properties (based on Abdel-Jawad and Al-Tabtabaei, 1999) and estimated chemical discharges of Doha West, Doha East and Shuwaikh. The estimate of antiscalant polymers is based on a typical dose of 2 ppm to the desalination feedwater. Copper loads are calculated from a brine contamination level of 15 ppb and chlorine discharges are based on a concentration of 250 ppb in the power plant cooling water and desalination discharge. Cooling water volumes depend on installed power generating capacity (Doha-West: 2,400 MW, Doha-East 1,050 MW, Shuwaikh 45 MW).

	Desalination feedwater [m^3]	**Capacity** m^3]	**Brine** [m^3]	**Desalination discharge** [m^3]	**Power plant cooling water** [m^3]	**Total** [m^3]
Doha-West	1,244,160	436,465	807,695	3,710,735	6,336,000	10,046,735
Doha-East	544,320	190,953	353,367	1,623,447	2,772,000	4,395,447
Shuwaikh	233,280	81,837	151,443	695,763	118,800	814,563
	2,021,760	709,256	1,312,505	6,029,944	9,226,800	15,256,744

	Polymer (2 ppm) [kg]	**Copper (15 ppb)** [kg]	**Chlorine (250 ppb)** [kg]
Doha-West	2,488	12.1	2,512
Doha-East	1,089	5.3	1,099
Shuwaikh	467	2.3	204
	4,044	19.7	3,814

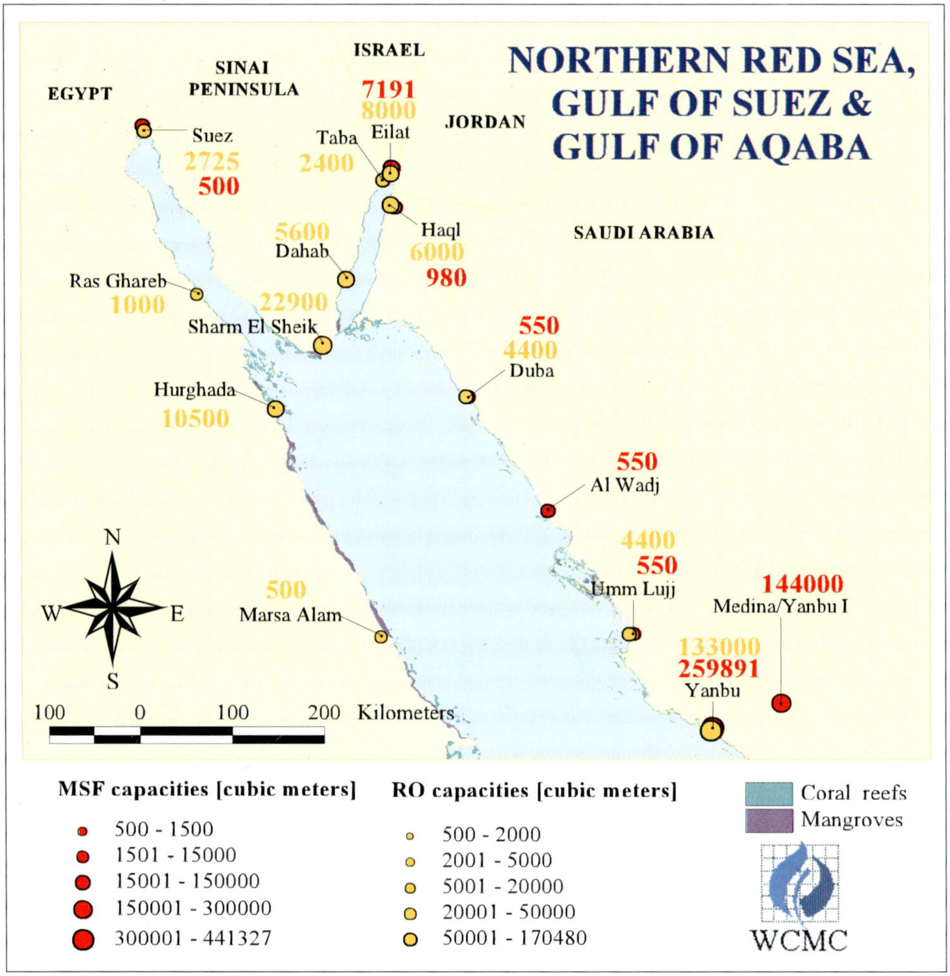

FIGURE A.8: Major MSF and RO seawater desalination plants in the northern Red Sea. Capacities based on Wangnick (1999). Data on coastline, corals and mangrove areas supplied by and copyright to World Conservation Monitoring Centre (WCMC) (2001). Saudi Arabian plants contribute more than 99 % and 85 % to total MSF and RO capacities. However, only 68 % of the Egyptian RO capacity could be identified and less than 1 % of the MSF seawater desalination capacity respectively. Major problems of identification are due to Arabian place-names in Wangnick (1999) that deviate from maps of the region. It should be noted that the unidentified plants may also be located on the Mediterranean shoreline of Egypt. Not included are brackish water desalination plants (e.g. 40,000 m^3 in Eilat, Israel), although they might discharge into the Red Sea, too.

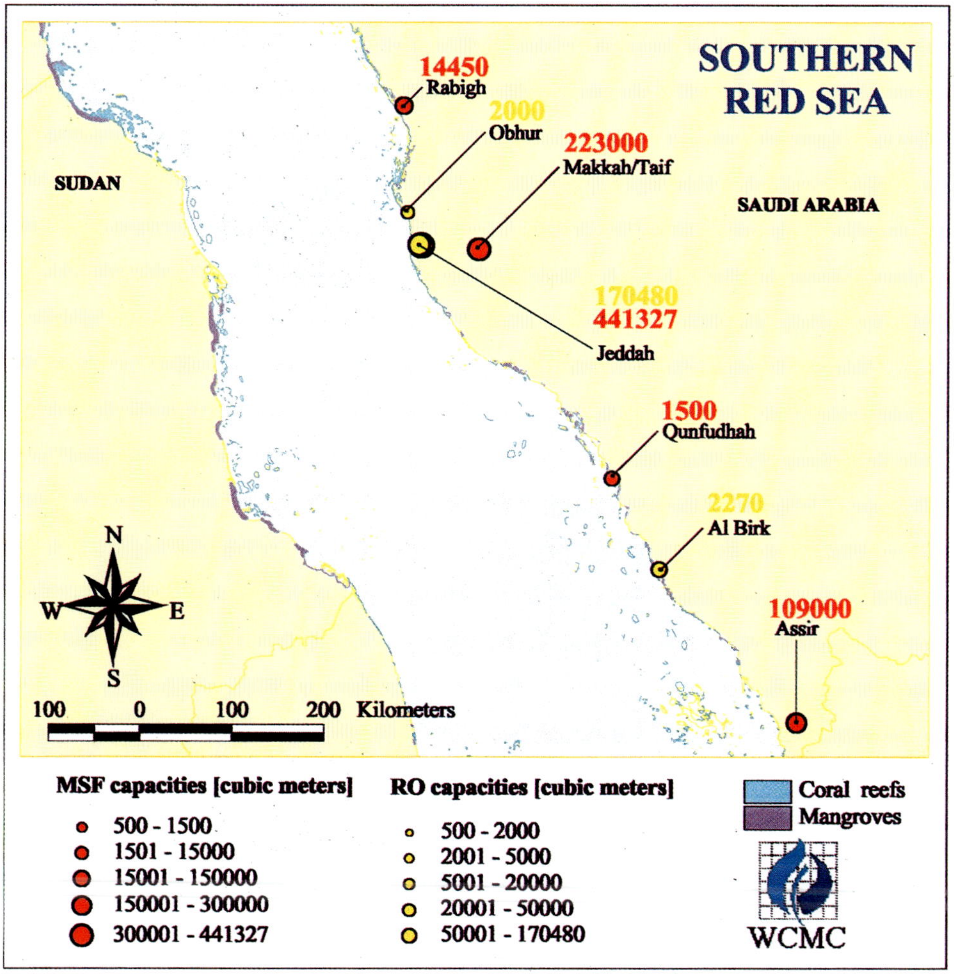

FIGURE A.9: Major MSF and RO seawater desalination plants in the southern Red Sea. Capacities based on Wangnick (1999). Data on coastline, corals and mangrove areas supplied by and copyright to World Conservation Monitoring Centre (WCMC) (2001).

Bibliography

ABART, E. (1993). Decision making strategy for handling biocides. Experience at Yuma Desalting Plant (Yuma AZ, USA). *Proceedings of IDA and WRPC World Conference on desalination and water treatment, Yokohama, Japan, 1993, In: Desalination*, **97**: 437–442.

ABDEL-JAWAD, M. and AL-TABTABAEI, M. (1999). Impact of current power generation and water desalination activities on Kuwaiti marine environment. *Proceedings of IDA World Congress on desalination and water reuse, San Diego, USA, September 1999*, **3**: 231–240.

ABDEL-MOATI, M. and KUREISHY, T. (1997). Dissolved copper, cadmium and lead in the coastal waters of Qatar, Arabian Gulf. *Indian Journal of Marine Sciences*, **26**: 143–149.

AGENDA 21 (1992). *Protection of the oceans, all kinds of seas, including enclosed and semi-enclosed seas, and coastal areas and the protection, rational use and development of their living resources.* published under http://www.un.org/esa/sustdev/agenda21.htm, Agenda 21, Chapter 17.

AHMADI, A. (1994). *Potential impacts of desalting on the environment.* In: Desalting as an environmentally friendly water treatment process, O.K. Buros (ED.), Seminar Summary Report, Water Treatment Technology Report No. 13, published under http://www.usbr.gov/water/media/pdfs/report013.pdf.

AHMED, M., SHAYYA, W., HOEY, D., MAHENDRAN, A., MORRIS, R. and AL-HANDALY, J. (2000). Use of evaporation ponds for brine disposal in desalination plants. *Desalination*, **130**: 155–168.

AL-AHMAD, M. and ALEEM, F. A. (1993). Scale formation and fouling problems effect on the performance of MSF and Ro desalination plants in Saudi Arabia. *Desalination*, **93**: 287–310.

AL-AWADHI, F. (1999). The year of the ocean and its crucial importance to the Gulf. *Desalination*, **123**: 127–133.

AL-AWADI, F. (1994). Environmental impacts of sites on P&D plants. *Desalination and Water Reuse*, **44**: 28–31.

AL-MAYOUF, A., AL-SUHYBANI, A. and AL-AMEERY, A. (1998). Corrosion inhibition of 304SS in sulfuric acid solutions by 2-methyl benzoazole derivates. *Desalination*, **116**: 25–33.

AL-MUTAZ, I. (1996). A comparative study of RO and MSF desalination plants. *Desalination*, **106**: 99–106.

AL-SHAMMIRI, M. and AL-DAWAS, M. (1997). Maxiumum recovery from seawater reverse osmosis plants in Kuwait. *Desalination*, **110**: 37–48.

AL-SHAMMIRI, M., SAFAR, M. and AL-DAWAS, M. (2000). Evaluation of two different antiscalants in real operation at the Doha research plant. *Desalination*, **128**: 1–16.

AL-TISAN, I., CHANDY, J., ABANMY, A. and HASSAN, A. (1995). Optimization of seawater reverse osmosis pretreatment: Part III - A microbiological approach. *Proceedings of IDA World Congress on desalination and water sciences Abu Dhabi, UAE, 1995*, **4**: 3–18.

ALAWADHI, A. (1997). Pretreatment plant design - Key to a successful reverse osmosis desalination plant. *Desalination*, **110**: 1–10.

ALI, M. and RILEY, J. (1986). The distribution of halomethanes in the coastal waters of Kuwait. *Marine Pollution Bulletin*, **17**: 409–414.

ALI, M. and RILEY, J. (1989). The production of brominated methanes in desalination plants in Kuwait. *Water Research*, **23**: 1099–1106.

ALTAYARAN, A. and MADANY, I. (1992). Impact of a desalination plant on the physical and chemical properties of seawater, Bahrain. *Water Research*, **26**: 435–441.

ANDERSON, L. and DYRSSEN, D. (1994). Alkalinity and total carbonate in the Arabian Sea. Carbonate depletion in the Red Sea and Persian Gulf. *Marine Chemistry*, **47**: 195–202.

ANDIJANI, I., AHMAD, S. and MALIK, A. (2000). Corrosion behavior of titanium metal in the presence of inhibited sulfuric acid at 50 °C. *Desalination*, **129**: 45–51.

BAIG, M. and KUTBI, A. A. A. (1998). Design features of a 20migd SWRO desalination plant, Al Jubail, Saudi Arabia. *Desalination*, **118**: 5–12.

BAKER, J. and DUDLEY, L. (1998). Biofouling in membrane systems - A review. *Desalination*, **118**: 81–90.

BASAHAM, A. S. and AL-LIHAIBI, S. (1993). Trace elements in sediments of the western Gulf. *In: Marine Pollution Bulletin, The 1991 Gulf War: Coastal and marine environmental consequences*, **27**: 103–107.

BASAHAM, A. S. and EL-SAYED, M. (1998). Distribution and phase association of some major and trace elements in the Arabian Gulf sediments. *Estuarine, Coastal and Shelf Science*, **46**: 185–194.

BASF (1994). *BASF speciality chemicals for water treatment and safety data sheet for Sokalan PM 10 I.* BASF Aktiengesellschaft, Marketing Spezialchemikalien, 67056 Ludwigshafen, Germany.

BETZDEARBORN (2000). *Sicherheitsdatenblätter.* BetzDearborn GmbH, Siemensring 44, 47877 Willich, Germany.

BOMM (2001). *pers. comm.* Versorgungsbetriebe Helgoland GmbH, 27498 Helgoland, Germany.

BOU-HAMAD, S., ABDEL-JAWAD, M., EBRAHIM, S., AL-MANSOUR, M. and AL-HIJJI, A. (1997). Performance evaluation of three different pretreatment systems for seawater reverse osmosis technique. *Desalination*, **110**: 85–92.

BURASHID, K. (1992). Bahrain's Seawater Intakes and Pretreatment Systems. *Desalination and Water Reuse*, **22**: 44–49.

BUROS, O. K. (1999). *The ABCs of desalting, 2nd edition.* FMC Corporation, published by the International Desalination Association (IDA), Topsfield, Massachusetts, USA and sponsored by the Saline Water Conversion Corporation (SWCC).

BUTT, F., RAHMAN, F. and BADURUTHAMAL, U. (1997). Evaluation of SHMP and advanced scale inhibitors for control of $CaSO_4$, $SrSO_4$, and $CaCO_3$ scales in RO desalination. *Desalination*, **109**: 323–332.

CHIDA, K. (1997). Reverse osmosis plants and maintenance experience in the Middle Eastern region. *Desalination*, **110**: 59–64.

CLARKE, M. and KEIJ, A. (1973). *Organisms as producers of carbonate sediment and indicators of environment in the southern Persian Gulf.* In: The Persian Gulf, Purser, B. and Seibold, E. (ED.), Springer-Verlag, New York, pp. 33-56.

COLES, S. and MCCAIN, J. (1990). Environmental factors affecting benthic infaunal communities of the western Arabian Gulf. *Marine Environmental Research*, **29**: 289–315.

DARTON, E. (1997). Scale inhibition techniques used in membrane systems. *Desalination*, **113**: 227–229.

DARWISH, M. (2000). *Co-generation power desalting plants.* Eurosummer School short course on sustainability assessment of water desalination technologies, Vilamoura, Portugal, November 2000.

DEL BENE, J., JIRKA, G. and LARGIER, J. (1994). Ocean brine disposal. *Desalination*, **97**: 365–372.

DUDLEY, L. and DARTON, E. (1997). Pretreatment procedures to control biogrowth and scale formation in membrane systems. *Desalination*, **110**: 11–20.

DUPONT (1994). *PERMASEP Products engineering manual, bulletin 1020, 2010 and 4010*. DuPont Company, Wilmington, DE 19898, USA.

EBRAHIM, S., ABDEL-JAWAD, M. and SAFAR, M. (1995). Conventional pretreatment system for the Doha Reverse Osmosis Plant: Technical and economic assessment. *Desalination*, **102**: 179–187.

EL-DAHAN, H. and HEGAZY, H. (2000). Gypsum scale control by phosphate ester. *Desalination*, **127**: 111–118.

EL-DESSOUKY, H. (2000). *Multi-stage flash desalination technologies.* Eurosummer School short course on sustainability assessment of water desalination technologies, Vilamoura, Portugal, November 2000.

EPA (1991). *Technical support document for water quality-based toxics control (TSP).* U.S. EPA, Office of Water, Washington, DC, EPA/505/2-90-001.

EPA (1998). *National recommended water quality criteria.* U.S. EPA, National Center for Environmental Publications and Information, available from 11029 Kenwood Road, Cincinnati, Ohio 45242, USA or http://www.epa.gov/ost.

EPA (1999). *Proposed national pollutant discharge elimination system (NPDES) general permit for reverse osmosis desalinization facilities in Saipan, Mariana Islands, Pacific Ocean.* Federal Register: April 1, 1999, volume 64, number 62, page 15749-15753.

EPA (2001). *Ocean discharge criteria: Revisions to ocean discharge criteria regulations.* U.S. EPA, Ariel Rios Building, Water Docket, Mail Code 4101, 1200 Pennsylvania Avenue, NW, Washington, DC 20460, published under http://www.epa.gov/owow/oceans/protecting-oceans/cwa403rule.pdf.

EPPLEY, R., RINGER, E. and WILLIAMS, P. (1976). Chlorine reactions with seawater constituents and the inhibition of photosynthesis of natural marine phytoplankton. *Est. Coastal Mar. Sci.*, **4**: 147.

EVANS, G., MURRAY, J., BIGGS, H., BATE, R. and BUSH, P. (1973). *The oceanography, ecology, sedimentology and geomorphology of the Trucial Coast barrier island complex, Persian Gulf.* In: The Persian Gulf, Purser, B. and Seibold, E. (ED.), Springer-Verlag, New York, pp. 233-277.

FALBE, J. and REGITZ, M. (Ed.) (1995). *Römpp Chemie Lexikon, 9th Edition on CD-ROM.* Georg Thieme Verlag, Stuttgart.

FENT, K. (1998). *Ökotoxikologie.* Georg Thieme Verlag, Stuttgart, 288p.

FILMTEC (2000). *FILMTEC membranes tech manual excerpts.* FilmTec Corporation, The Dow Chemical Company, published under http://www.dow.com/liquidseps.

FINAN, M., SMITH, S., EVANS, C. and MUIR, J. (1989). Belgard EV - 15 years' experience in scale control. *Desalination,* **73**: 341–357.

FISIA ITALIMPIANTI (2001). *Desalination Taweelah Abu Dhabi.* Plant supplier of the Taweelah B plant in Abu Dhabi, published under http://www.fisiait.com/activities/desalination.

FMC (1994). *Flocon 100 general product information.* FMC Corporation, Process Additives Division, Tenax Road, Trafford Park, Manchester M17 1WT.

FMC (1997). *Belgard EV2030.* FMC Corporation, published under http://www.desalination.com/FMC/EV2030.html.

FOWLER, S., READMAN, J., OREGIONI, B., VILLENEUVE, J.-P. and MCKAY, K. (1993). Petroleum hydrocarbons and trace metals in nearshore Gulf sediments and biota before and after the 1991 war: An assessment of temporal and spatial trends. *In: Marine Pollution Bulletin, The 1991 Gulf War: Coastal and marine environmental consequences,* **27**: 171–182.

GESAMP (1986). *Environmental Capacity. An approach to marine pollution prevention.* IMO/FAO/Unesco/WMO/WHO/IAEA/UN/UNEP Joint Group of Experts on the Scientific Aspects of Marine Pollution, GESAMP Reports and Studies No. 30.

GILL, J. (1999). A novel inhibitor for scale control in water desalination. *Desalination,* **124**: 43–50.

GITA SHENASI CARTOGRAPHIC & GEOGRAPHIC INSTITUTE (undated). *The political & historical map of Persian Gulf.* No.15, Arfa Street, P.O. Box 14155-3441, Tehran, Iran.

GKSS (1986). *The influence of yellow substances on remote sensing of seawater constituents from space.* GKSS research centre Geesthacht, Volume I Summary Report.

GLEICK, P. (1999). The Human Right to Water. *Water Policy,* **1**: 487–503.

GLUCINA, K. and LAÎNÉ, J.-M. (2000). *Reverse osmosis water desalination technologies II.* Eurosummer School short course on sustainability assessment of water desalination technologies, Vilamoura, Portugal, November 2000.

GRAVES, M., BLACK, B. and DODSON, J. (2001). *Desalination for Texas water supply.* HDR Engineering Inc. and Nueces River Authority, published under http://www.tamuk.edu/STEC2000/STEC2000present/Graves_Black.html.

HAMED, O., AL-SOFI, M., IMAM, M., MARDOUF, K., AL-MOBAYED, A. and EHSAN, A. (2000). Evaluation of polyphosphonate antiscalant at a low dose rate in the Al-Jubail Phase II MSF plant, Saudi Arabia. *Desalination*, **128**: 275–280.

HAMED, O., AL-SOFI, M., MUSTAFA, G. and DALVI, A. (1999). The performance of different anti-scalants in multi-stage flash distillers. *Desalination*, **123**: 185–194.

HASSAN, A., AL-SOFI, A., AL-AMOUDI, A., JAMALUDDIN, A., FAROOQUE, A., ROWAILI, A., DALVI, A., KITHER, N., MUSTAFA, G. and AL-TISAN, I. (1998). A new approach to membrane and thermal seawater desalination processes using nanofiltration membranes (Part 1). *Desalination*, **118**: 35–51.

HASSON, D., SEMIAT, R., BRAMSON, D., BUSCH, M. and LIMONI-RELIS, B. (1998). Suppression of $CaCO_3$ scale deposition by antiscalants. *Desalination*, **118**: 285–296.

HEITMANN, H.-G. (1990). *Saline water processing.* VCH Verlagsgesellschaft, Weinheim, 332p.

HERBRANSON, L. and SUEMOTO, S. (1993). Desalination research - Current needs and approaches - A U.S. perspective. *Proceedings of IDA and WRPC World Conference on desalination and water treatment, Yokohama, Japan, 1993*, **1**: 209–216.

HINRICHSEN, D. (1998). *Coastal waters of the world. Trends, threats, and strategies.* Island Press, Washington, D.C., 275p.

HOEPNER, T. (1999). A procedure for environmental impact assessment (EIA) for desalination plants. *Desalination*, **124**: 1–12.

HÖPNER, T. and WINDELBERG, J. (1996). Elements of environmental impact studies on coastal desalination plants. *Desalination*, **108**: 11–18.

HSDB (2001). *Hazardous Substances Databank (HSDB), a database of the National Library of Medicine's TOXNET system.* Full records a – nn, which are quoted in the text, are listed on page 132. Access to database under http://toxnet.nlm.nih.gov.

HUSSAIN, A. and AHMED, A. (1998). The Addur SWRO desalination plant towards a full plant production. *Desalination*, **120**: 143–152.

HYDRANAUTICS (2000). *Technical service bulletins.* Hydranautics, published under http://www.membranes.com.

ICSC (2000). *International chemical safety cards.* International Programme on Chemical Safety carried out by the World Health Organization (WHO), the International Labor Organization (ILO) and the environmental programme of the United Nations (UNEP), published under http://www.cdc.gov/niosh/ipcs/icstart.html.

IMAM, M., AL-RASHEED, R., AL-TISAN, I., ABDULGADER, A., MUSTAFA, G., AL-MOBAYED, A. S., EHSAN, A. and BROSE, D. (2000). Studies on a polyoxypropylene glycol-based antifoaming agent in MSF plants. *Desalination*, **129**: 187–196.

ISO, S., SUIZU, S. and MAEJIMA, A. (1993). The lethal effect of hypertonic solution and avoidance of marine organisms in relation to the dscharged brine from desalination plant. *Proceedings of IDA and WRPC World Conference on desalination and water treatment, Yokohama, Japan, 1993*, **2**: 109–120.

JIRKA, G., DONEKER, R. and HINTON, S. (1996). *User's manual for cormix: A hydrodynamic mixing zone model and decision support system for pollutant discharges into surface waters.* DeFrees Hydraulics Laboratory, School of Civil and Environmental Engineering, Cornell University, Ithaca, New York 14853-3501, published under http.steens.ese.ogi.edu.

KENNISH, M. (1997). *Practical handbook of estuarine and marine pollution, 2nd edition.* CRC Press, Boca Raton, 524p.

KERSTEN, M. and KRÖNCKE, I. (1990). *Schwermetalle in den Sedimenten.* In: Warnsignale aus der Nordsee, Lozan, J. L., Lenz, W. and Rachor, E. (ED.), Parey Buchverlag, Hamburg, pp. 41-47.

KHORDAGUI, H. (1992). Conceptual approach to selection of a control measure for residual chlorine discharge in Kuwait Bay. *Environmental Management 16*, **3**: 309–316.

KHORDAGUI, H. (1997). *Environmental aspects of brine reject from desalination industry in ESCWA region.* ESCWA, Beirut, Lebanon, 1997.

KIMES, J. (1995). The regulation of concentrate disposal in Florida. *Desalination*, **102**: 82–92.

KNOLL, W. (2001). *pers. comm. on Cobratec CO-99 material safety data sheet information.* Technical Service PMC, Ludwigshafen, Germany, wknoll@raschig.de.

KUREISHY, T. (1993). Concentrations of heavy metals in marine organisms around Qatar before and after the Gulf war oil spill. *In: Marine Pollution Bulletin, The 1991 Gulf War: Coastal and marine environmental consequences*, **27**: 183–186.

KUWAIT REGIONAL CONVENTION (1978). *Kuwait Regional Convention for co-operation on the protection of the marine environment from Pollution, Kuwait, 24 April 1978.* published under http://sedac.ciesin.org/pidb/texts/kuwait.marine.pollution.1978.html.

LUU, C. (1994). Improved chelators and sequestrants for Army Reverse Osmosis Water Purification Units (ROWPUs). *Desalination*, **97**: 165–170.

MABROOK, B. (1994). Environmental impact of waste brine disposal of desalination plants, Red Sea, Egypt. *Desalination*, **97**: 453–465.

MALIK, A., AHMAD, S., ANDIJANI, I. and AL-FOUZAN, S. (1999). Corrosion behavior of steels in Gulf seawater environment. *Desalination*, **123**: 205–213.

MANNAA, A. (1994). Environmental impacts of dual purpose plants. *Desalination and Water Reuse*, **41**: 46–49.

MATTICE, J. and ZITTEL, H. (1976). Site-specific evaluation of power plant chlorination. *J. Water Pollut. Control Fed*, **48**: 2284.

MEDRC (2001). *Saudi Arabia: the growing need for investment in desalination.* Watermark, the Newsletter of the Middle East Desalination Reseach Center (MEDRC), Issue 12, P.O Box 21, P.C. 133, Oman, www.medcr.org.

MICKLEY, M. (1995). Environmental considerations for the disposal of desalination concentrates. *Desalination and Water Reuse*, **54**: 56–61.

MORGAN, P. (1994). *SW_DENS0 Denisty of sea water at atmospheric pressure.* A Matlab-based shareware for the calculation of seawater density, based on Unesco (1983): Algorithms for computation of fundamental properties of seawater, Unesco Tech. Pap. in Mar. Sci., 44: 53.

MORGAN, P. (1998). *SW_SATO2 Saturation of O2 in sea water.* A Matlab-based shareware for the calculation of O_2 saturation, based on Weiss, R. F. (1970): The solubility of nitrogen, oxygen and argon in water and seawater, Deap-Sea Research, 17: 721-735.

MORTON, A., CALLISTER, I. and WADE, N. (1996). Environmental impacts of seawater distillation and reverse osmosis processes. *Desalination*, **108**: 1–10.

MUBARAK, A. (1998). A kinetic model for scale formation in MSF desalination plants. Effects of antiscalants. *Desalination*, **120**: 33–39.

NING, R. (1999). Reverse osmosis process chemistry relevant to the Gulf. *Desalination*, **123**: 157–164.

OBAID, M. and HAMIDA, A. (1998). Practical solutions to problems experienced in open seawater RO plants operating on the Arabian Gulf. *Desalination*, **120**: 137–142.

OLDFIELD, J. and TODD, B. (1996). Environmental aspects of corrosion in MSF and RO desalination plants. *Desalination*, **108**: 27–36.

PANTELL, S. (1993). *Seawater Desalination in California.* California Coastal Commission, published under http://www.coastal.ca.gov/desalrpt/dchap3.html.

PAQUIN, P., SANTORE, R., ANID, P. and TORO, D. D. (1999). *Revisiting the aquatic impacts of copper from water-cooled copper alloy condenser discharges.* Proceedings of IDA World Congress on desalination and water reuse, San Diego, USA, September 1999.

PATEL, S. and FINAN, M. (1999). New antifoulants for deposit control in MSF and MED plants. *Desalination*, **124**: 63–74.

PMC (2000). *Cobratec.* PMC Specialities Group, Inc., published under http://www.metalprotection.com.

Price, A. (1993). The Gulf: Human impacts and management initiatives. *In: Marine Pollution Bulletin, The 1991 Gulf War: Coastal and marine environmental consequences*, **27**: 17–27.

Purser, B. and Seibold, E. (1973). *The principal environmental factors influencing holocene sedimentation and diagenesis in the Persian Gulf.* In: The Persian Gulf, Purser, B. and Seibold, E. (ED.), Springer-Verlag, New York, pp. 1-9.

Redondo, J. and Lomax, I. (1997). Experiences with the pretreatment of raw water with high fouling potential for reverse osmosis plant using FILMTEC membranes. *Desalination*, **110**: 167–182.

Redox-Chemicals (2001). *Material safety data sheets.* Redox Chemicals PTY LTD, Australia, published under http://www.redox.com.au/msds/msds.data/POMAAC.htm.

Reynolds, M. (1993). Physical oceanography of the Gulf, Straits of Hormuz, and the Gulf of Oman: Results from the Mt. Mitchell expedition. *In: Marine Pollution Bulletin, The 1991 Gulf War: Coastal and marine environmental consequences*, **27**: 35–59.

Ribiero, J. and Spiewak, I. (1998). *Desalination research priorities - A European survey, IPTS EUR 18087 EN.* Institute for Prospective Technological Studies, Isla de la Cartuja, 41092 Sevilla, Spain, 36p.

Saad, M. (1985). Distribution of nutrients in Shatt-Al-Arab estuary. 2. Nutrient salts. *Bulletin Institute Oceanogr. Fish. Cairo*, **11**: 27–42.

Sadiq, M. (1996). Metal bioaccumulation by sea urchin (*Echinometra mathaei*) from the Saudi coastal areas of the Arabian Gulf: 2. cadmium, copper, chromium, barium, calcium, and strontium. *Bulletin of environmental contamination and toxicology*, **57**: 964–971.

Saeed, M., Jamaluddin, A., Tisan, I., Lawrence, D., Al-Amri, M. and Chida, K. (2000). Biofouling in a seawater reverse osmosis plant on the Red Sea coast, Saudi Arabia. *Desalination*, **128**: 177–190.

Saeed, T., Khordagui, H. and Al-Hashash, H. (1999). Contribution of power/desalination plants to the levels of halogenated volatile liquid hydrocarbons in the coastal areas of Kuwait. *Desalination*, **121**: 49–63.

Saline Water Conversion Corporation (1995). *Report No. SWCC(RDC)-36.* Saline Water Conversion Corporation, Saudi Arabia, March 1995.

SCOTT, D. (1995). *A directory of wetlands in the Middle East.* IUCN, Gland, Switzerland and IWRB, Slimbridge, U.K., 560pp, 13 maps, published under http://www.wetlands.agro.nl/wetland_inventory/middle_east_review.html.

SCOTT, G., SAMMONS, T., MIDDAUGH, D. and HEMMER, M. (Ed.) (1982). *In: Physiological mechanisms of marine pollutant toxicity.* Academic Press, New York.

SHAMS EL DIN, A., ARAIN, R. and HAMMOUD, A. (2000a). On the chlorination of seawater. *Desalination*, **129**: 53–62.

SHAMS EL DIN, A., AZIZ, S. and MAKKAWI, B. (1994). Electricity and water production in the Emirate of Abu Dhabi and its impact on the environment. *Desalination*, **97**: 373–388.

SHAMS EL DIN, A., EL-DAHSHAN, M. and TAJ EL DIN, A. (2000b). Dissolution of copper and copper-nickel alloys in aerated dilute HCL solution. *Desalination*, **130**: 89–97.

SHAMS EL DIN, A. and MAKKAWI, B. (1998). Operation processes affecting corrosion in MSF distillers. *Desalination*, **115**: 33–37.

SHAMS EL DIN, A. and MOHAMMED, R. (1997). Physico-chemical studies on BELGARD EV and BELGARD EV 2000, Part III: Analytical procedures for the estimation of active components. *Desalination*, **114**: 77–84.

SHAMS EL DIN, A. and MOHAMMED, R. (1998). Kinetics of reaction between hydrogen peroxide and hypochlorite. *Desalination*, **115**: 145–153.

SHEIKHOLESLAMI, R. (1999). Fouling mitigation in membrane processes. *Desalination*, **123**: 45–53.

SHEPPARD, C. (1993). Physical environment of the Gulf relevant to marine pollution: An overview. *In: Marine Pollution Bulletin, The 1991 Gulf War: Coastal and marine environmental consequences*, **27**: 3–8.

SHEPPARD, C., PRICE, A. and ROBERTS, C. (1992). *Marine ecology of the Arabian Region.* Academic Press, London, 359p.

SHRIADAH, M. M. A. (1999). Heavy metals in mangrove sediments of the United Arab Emirates shoreline (Arabian Gulf). *Water, Air, and Soil Pollution*, **116**: 523–534.

SPIEGLER, K. and EL-SAYED, Y. (1994). *A desalination primer - Introductory book for students and newcomers to desalination.* Balaban Desalination Publications, School for Scientific Communication, Mario Negri Sud Research Institute, Italy, 215p.

SUBBA RAO, D. and AL-YAMANI, F. (2000). The Arabian Gulf. *In: Seas at the millenium: An environmental evaluation, chapter 53*, **2**: 1–16.

TODD, B. (2001). *Copper-nickel alloys in desalination systems.* Marine applications of copper-nickel alloys, Section 4, Nickel Development Institute, published under http://marine.copper.org/.

TORAY (1989). *Romembra operation, maintenance and handling manual.* Toray Industries, Inc., Membrane Products Dept., Toray Building, 8-1, Mihama 1-chome, Urayasu, Chiba 279-8555, Japan, published under http://www.dow.com/liquidseps.

TRISEP (1999). *Miscellaneous Chemicals.* Trisep Corporation, published under http://www.trisep.com.

UK HYDROGRAPHIC OFFICE (1990). *Admirality chart no. 1214, Khalij al Kuwait.* Admirality Way, Taunton, Somerset, TA1 2DN, United Kingdom.

UNEP (1997). *Assessment of land-based sources and activities affecting the marine environment in the Red Sea and Gulf of Aden, UNEP Regional Seas Reports and Studies No. 166.* UNEP/ GPA Coordination Office, P.O. Box 16227 2500 BE The Hague, The Netherlands, published under http://www.gpa.unep.org/documents/technical/rseas_reports/default.htm.

UNEP (1999). *Overview on land-based sources and activities affecting the marine environment in the ROPME sea area, UNEP Regional Seas Reports and Studies No. 168.* UNEP/ GPA Coordination Office, P.O. Box 16227 2500 BE The Hague, The Netherlands, published under http://www.gpa.unep.org/documents/technical/rseas_reports/default.htm.

WANGNICK, K. (1999). *IDA Worldwide Desalting Plants Inventory Report No. 16.* Wangnick Consulting GmbH, 27442 Gnarrenburg, Germany, 247p.

WILCOCK, J. and FINAN, M. (1994). Ensuring the safety of additives used in desalination plants to produce potable water: The role of independent approval agencies and comprehensive toxicological and ecotoxical product profiles. *Desalination,* **97**: 443–451.

WILF, M. and KLINKO, K. (1998). Effective new pretreatment for seawater reverse osmosis systems. *Desalination,* **117**: 323–331.

WINTERS, H. (1995). Biofouling - Its history and how it affects today's desalination industry. *Proceedings of IDA World Congress on desalination and water sciences, Abu Dhabi, UAE, 1995,* **1**: 255–264.

WINTERS, H. (1997). Twenty years experience in seawater reverse osmosis and how chemicals in pretreatment affect fouling of membranes. *Desalination,* **110**: 93–96.

WOODWARD CLYDE CONSULTANTS (1991). *Final environmental impact report (FEIR).* City of Santa Barbara and Ionics, Incorporated's Temporary Emergency Desalination Project, SB-106-90, State Clearing House No. 9010859.

WORLD CONSERVATION MONITORING CENTRE (WCMC) (2001). *GIS DATA from the Arabian Gulf and Red Sea.* 219 Huntingdon Road, Cambridge CB3 0DL, United Kingdom, e-mail: info@wcmc.org.uk, http://www.wcmc.org.uk.

Records from the Hazardous Substances Databank (HSDB)

a Ellgaard E.G. and Gilmore J.Y. Jr, J Fish Biol 25 (2): 133-7 (1984);

b Environment Canada, Tech Info for Problem Spills: Hydrochloric acid (Draft) p.87 (1981);

c U.S. Coast Guard, Department of Transportation. CHRIS - Hazardous Chemical Data. Volume II. Washington, D.C., U.S. Government Printing Office, 1984-5;

d Environment Canada, Tech Info for Problem Spills: Chlorine (Draft) p.75-81 (1981);

e Sanders J.G., Environ Sci Technol 18 (5): 383-5 (1984);

f Mattice J.S. et al, Water Res 15 (7): 923-8 (1981);

g Itoh N. and Shinya M., Marine Chemistry 45: 95-103 (1994);

h Tabak H.H. et al, J Water Pollut Cont Fed 53: 1503-18 (1981);

i Meylan W.M. and Howard P.H., J Pharm Sci 84: 83-92 (1995);

j USEPA Ambient Water Quality Criteria Doc, Halomethanes p.B-5 (1980) EPA 440/5-80-051;

k Verschueren, K., Handbook of Environmental Data on Organic Chemicals, 3rd ed., New York, Van Nostrand Reinhold Co., 1996, 333p;

l U.S. Environmental Protection Agency's Integrated Risk Information System (IRIS) on Dibromochloromethane (124-48-1) Available from: http://www.epa.gov/ngispgm3/iris on the Substance File List as of March 15, 2000;

m USEPA Ambient Water Quality Criteria Doc, Chlorine p.7-8 (1984) EPA 440-5-84-030;

n Seegert et al, The effects of monochloramine on selected riverine fishes, Trans Am Fish Soc 108(1) 88 (1979);

o Budavari, S. (ed.), The Merck Index - An Encyclopedia of Chemicals, Drugs, and Biologicals, Whitehouse Station, NJ, Merck and Co. Inc., 1996, 1481p;

p USEPA Ambient Water Quality Criteria Doc, Copper p.B-14 (1980) EPA 440/5-80-036;

q Seiler, H.G., Sigel H. and Sigel A. (eds.), Handbook on the Toxicity of Inorganic Compounds, New York, Marcel Dekker Inc., 1988, 273p;

r Callahan, M.A., Slimak M.W. and N.W. Gabel, Water-Related Environmental Fate of 129 Priority Pollutants, Volume I, EPA-440/4 79-029a, Washington DC, U.S. Environmental Protection Agency, December 1979;

s Rosko J.J. and Rachlin J.W., Bull Torrey Bot Club 102: 100 (1975);

t Okazaki R.K., Bull Environ Contam Toxicol 16: 658-64 (1976);

u Olsen K.R. and Harrel R.C., Contrib Mar Sci 17: 9-13 (1973);

v Dave G., Comp Biochem Physiol, Comp Pharmacol Toxicol 78C (2): 439-43 (1984);

w Gould E. et al, Mar Biol (Berlin) 97 (2): 217-23 (1988);

x National Research Council Canada, Effects of Nickel in the Canadian Environment (1981) NRCC 18568;

y Sunderman F.W. Jr and Oskarsson A., in: Metals and Their Compounds in the Environment, pp. 1101-1126, Merian E. ed., NY, Weinheim (1991);

z Anderson P.R. and Christensen T.H., J Soil Sci 39: 15-22 (1988);

aa Mopper K. and Stahovec W.L., Marine Chem 19: 305-21(1986);

bb Chameides W.L. and Davis D.D., Nature 304: 427-9 (1983);

cc Lyman W.J. et al, Handbook of Chemical Property Estimation Methods, Washington DC, Amer Chem Soc, (1990);

dd Kubodera T. et al, Yakugaku 27 (12): 838-42 (1978);

ee Huddleston R.L. and Nielsen A.M., Soap Cosmet Chem Spec 55: 36-8 (1979);

ff Rott B. et al, Chemosphere 11: 531-8 (1982);

gg Cook T.M. and Goldman C.K., Degradation of Anionic Detergents in Chesapeake Bay, Chesapeake Sci 15(1): 52-55 (1974);

hh Anderson D.J. et al, Appl Environ Microbiol 56: 758-63 (1990);

ii Boatman R.J. et al, Environ Tox Chem 5: 233 (1986);

jj Belly R.T. et al, Appl Microbiol 29: 787 (1975);

kk Tiedje J.M., J Environ Qual 6: 21 (1977);

ll Gardiner J., Water Res 10: 507 (1976);

mm Bishop W.E. and Maki A.W., in: Aquatic Toxicology, pp 61-77, Eaton J.G. ed., ASTM STP 707 (1980);

nn Mann H., Arch Fischereiwiss 24 (1-3): 171-5 (1973)

Index

Bold page numbers refer to Figures.

Q

R

S

T